Security of Biochip Cyberphysical Systems

Shayan Mohammed • Sukanta Bhattacharjee
Yong-Ak Song • Krishnendu Chakrabarty
Ramesh Karri

Security of Biochip Cyberphysical Systems

 Springer

Shayan Mohammed
The University of Texas at Dallas
Richardson, TX, USA

Yong-Ak Song
New York University Abu Dhabi
Abu Dhabi, United Arab Emirates

Ramesh Karri
New York University
Brooklyn, NY, USA

Sukanta Bhattacharjee
Indian Institute of Technology Guwahati
Guwahati, India

Krishnendu Chakrabarty
Duke University
Durham, NC, USA

ISBN 978-3-030-93276-3   ISBN 978-3-030-93274-9   (eBook)
https://doi.org/10.1007/978-3-030-93274-9

This Springer imprint is published by the registered company Springer Nature Switzerland AG
The registered company address is: Gewerbestrasse 11, 6330 Cham, Switzerland

Acknowledgment

The authors would like to thank Jack Tang (New York University), Robert Willie (Johannes Kepler University Linz), Tung-Che Liang (Duke University), and Ajymurat Orozaliev (New York University Abu Dhabi) for their valuable inputs to our work.

Contents

List of Figures

List of Tables

Chapter 1
Introduction

Biochemistry deals with chemical reactions in a variety of life forms. It is the basis for practical advances in medicine, agriculture, molecular genetics, and biotechnology. Traditionally, biochemical protocols are carried out in wet labs, wherein qualified professionals manually perform a sequence of steps using specialized instruments. Often the ability to perform a biochemical protocol is limited by either monetary/time cost or availability of specialized professionals/machinery. Automation can help overcome the latter limitation. The cost of experiments is a function of the quantity of materials used and the experiment duration. So, if the experiments can be done on a miniaturized level, the cost can be minimized. A vision of expanding the biochemical applications motivated the idea of a lab-on-a-chip (LoC)—*shrinking a wet lab onto a tiny platform.*

Microfluidic technologies are one of the major driving forces towards the miniaturization of laboratory-based biochemical protocols. A microfluidic biochip or lab-on-a-chip (LoC) performs biochemical reactions by consuming nano-/pico-liter volume of reagents [134]. These platforms provide advantages such as minimal sample and reagent use, quicker results, automation, and reduced reliance on high-skilled personnel.

Biochips have brought a complete paradigm shift in several biochemical applications, some of which are referred to in this book. These chips have made a profound impact on health care by redefining point-of-care diagnostics [45], drug research and development [77], as well as personalized medicine [128]. The biochips make diagnostics affordable and accessible compared to traditional bio-lab-based platforms: For example, the immunoassay platform shown in Fig. 1.1 performs low-cost detection of measles and rubella viruses using a single drop of blood. This was deployed in refugee camps where many basic life necessities were inaccessible [9, 60]. Further, biochips enable diagnostics that were not possible in traditional labs: For example, microfluidics enables the development of personalized medicine needed for cancer patients by running thousands of parallel tests on patient's bio-sample. Using traditional techniques, it is too time-consuming to

S. Mohammed et al., *Security of Biochip Cyberphysical Systems*,
https://doi.org/10.1007/978-3-030-93274-9_1

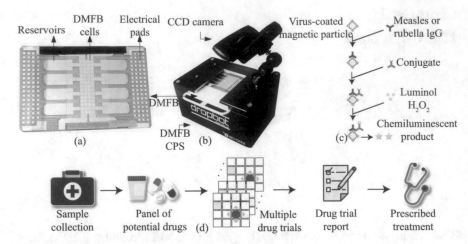

Fig. 1.1 An open-source DMFB system: (**a**) DMF biochip and (**b**) DropBot platform (source: https://sci-bots.com/). (**c**) Rubella/measles immunoassay description [100]. (**d**) A personalized drug development process for cancer patients [7, 24]

exactly determine which of the many clinical trials that are on offer is appropriate for a particular patient [7].

The global biochip market is projected to reach from $5.7 billion in 2018 to $12.3 billion by 2025 [28]. This is corroborated by the sales [25], investment [21], and acquisitions [15] reported by microfluidic companies. Baebies' SEEKER, a DMFB-based immunoassay platform, received FDA approval in 2016 [6]. Since then, Baebies has shipped three million tests and raised $13 million in funding for product development [25]. The SEEKER provides a high-throughput quantitative measurement of deadly diseases from the dried blood spot of newborns. Fluidigm is another leading microfluidic platform provider with a focus on cancer, immunology, and immunotherapy [13]. Their total revenue reported in 2019 was $30.1 million [22]. 10x Genomics uses a combination of microfluidics, chemistry, and bioinformatics for single-cell analysis. Since its founding in 2012, it has received $243 million in funding until 2018 [21].

1.1 Threat to Integrity

As biochips are penetrating the market, security and trust issues are being uncovered. Biochips have multiple usage scenarios, such as in a biomedical research lab and in a remote location. Depending on the usage scenario, biochemical protocol implementation faces different threats. To highlight this, we describe three real-life situations:

1. A disgruntled employee can tamper with the biochemical experiments to take revenge on colleagues or management [1]. In 2020, a chemist at a water treatment plant was found guilty of tampering with a colleague's water test for months [23]. The usage of biochips in such labs increases the risk of such attacks due to the biochip's easy controllability.
2. An unfaithful biochip designer, who uses fraudulent or falsified claims, is a threat to the users, investors, and regulators. *Edison* microfluidic blood testing device from *Theranos* faced technical, commercial, and legal challenges over the scientific basis of its technologies [27]. Such incidents gather a lot of negative press and hamper the progress in such technologies [26].
3. Studies have flagged security flaws in medical devices such as tampering of controls, denial-of-service, data theft, and ransom attacks [2]. This has led to a recall of a large number of medical devices and a re-evaluation of their regulations [20]. A biochip cyber-physical system (CPS) is similar to the current medical devices, which consists of hardware, software, and network connections [54]. As biochips are becoming an integral part of health care services, these threats become more pronounced.

These threats may lead to a loss of revenue and trust or, more importantly, jeopardizes the well-being of its users [123]. They can cause denial-of-service and wrong bioassay outcomes. Addressing these threats becomes even more critical as the biochips are being used in artificial-intelligence-based decision making [109], and the emergence of miniaturized versions of oneself for medical tests [139].

1.2   The Threat to IP Rights

Companies, such as pharmaceuticals, invest large sums of money and person-hours in a slow and expensive bio-protocol development process laced with tough regulations. For example, a DMFB implementation of a thyroid-stimulating hormone immunoassay requires hundreds of experiments to determine the right bio-protocol parameters [46]. This process is prone to the stealing of sensitive research data [5]. In 2016, two scientists at a leading pharmaceutical company were indicted for conspiring with a competitor to steal promising drug research secrets [4]. For rapid and low-cost drug development, pharmaceutical companies are using various types of microfluidic biochips that minimize the assay time and reagent requirement [77]. This opens new avenues for IP theft. Biochip applications inside and outside a lab pose a new challenge to IP protection:

Inside a Lab Due to the transparent nature of bioprotocol implementation on the biochip, the bioprotocol sequence can be reverse-engineered. This can be achieved using the biochip snapshots and/or the controlling actuation sequence [42]. This opens the door for bio-IP theft by reverse-engineering.

Outside a Lab Traditionally, bio-protocols were implemented in controlled laboratory environments. Biochip technology permits the execution of bio-protocols in remote settings on a miniaturized platform. Though this enables new applications such as point-of-care diagnostics [6], it makes the bio-protocols susceptible to illegal copy and counterfeit production.

1.3 Book's Scope and Road Map

Biochip CPS is targeted for safety-critical applications such as diagnostics and drug development. Therefore, security is of paramount importance. Due to the multidisciplinary nature of the systems, overlapping the fields of computer science, microfluidics, and biochemistry, it requires a holistic approach to assess the security of biochip CPS. This book uncovers new potential threats and trust issues, which must be studied as this emerging technology is poised to be adopted at a large scale. The results emerging from this book answer the following crucial questions: (1) how to secure biochip CPS by leveraging the available resources in different application contexts? (2) how to ensure intellectual property (IP) is protected against theft and counterfeits? This book aids secure biochip CPS design by helping bridge the knowledge gap at the intersection of multi-disciplinary technology that drives biochip CPS.

   Rest of this chapter layouts the requisite background for understanding the biochip design. In Chap. 2 various threats to biochip design flow are discussed with case studies. The first step towards secure biochip design is choosing an appropriate architecture. In Chap. 3, we show how to perform a security assessment of a given architecture. Next, a biochip designer needs to be equipped with CAD tools that can (1) verify a design against security properties and (2) enforce the security properties during run-time. Chapter 4 delves into the development of security analysis tools: verification of checkpoint and ML-based attack detection. Further, we focus on the protection of bio-IP in the last two chapters. Watermarking of bio-IP is described in Chap. 5, and the obfuscation of bio-IP is described in Chap. 6. Most of the textbook focuses on digital microfluidic devices for consistency. However, the concepts and results are also applicable in the context of continuous-flow microfluidic devices.

1.4 Biochip Systems

A biochip platform enables microfluidic operations such as dispensing, mixing, and splitting. These, in turn, can be used to build more sophisticated protocols for biochemical analysis [56]. Several biochip platforms have been proposed, such as digital microfluidic biochip (DMFB) or continuous-flow-based microfluidic biochip (CFMB). CFMBs manipulate fluid flow through a network of microchannels by actuating pressure-driven microvalves [105]. DMFB offers a programmable

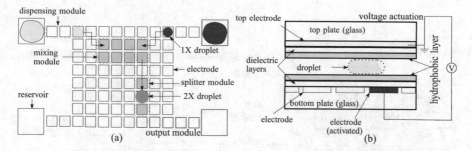

Fig. 1.2 DMFB schematic: (**a**) top and (**b**) cross-sectional view

fluidic platform in which discrete fluid droplets are manipulated through electrical actuations [132]. These are briefly described in the subsequent subsections.

1.4.1   Digital Microfluidic Biochips

DMFBs manipulate fluids in discrete quantities and, in recent years, have become synonymous with biochips based on the electrowetting-on-dielectric (EWOD) principle: the control of the contact angle between a droplet and substrate through the application of a suitable electric potential [70]. The DMFB structure is typically composed of two overlapping plates separated by a gap (Fig. 1.2a,b). The bottom plate has a two-dimensional array of electrodes, while the top plate acts as a ground electrode. Voltages are applied to the electrodes to manipulate the wetting forces on the droplet. Droplets are attracted to the neighboring electrodes with higher voltages [82]. This results in the controlled movement of droplets in the horizontal and vertical directions on the biochip. Operations such as dispensing, transporting, mixing, and splitting can be carried out by applying actuation sequences: control signals that are correctly timed and sequenced. Based on these fundamental operations, bioassays for clinical and DNA sequencing can be implemented [29, 57].

1.4.2   Continuous-Flow Microfluidic Biochips

CFMB consists of two layers of permanently etched microchannels called the flow and the control layer, as shown in Fig. 1.3a. At the intersection of the two layers, a "valve" is formed that can be controlled by an external pressure source. When the valve is pressurized, the flexible membrane of the control layer deflects deep into the flow layer blocking the fluid flow (ref. Fig. 1.3b). By controlling the opening/closing of the valves, complex fluid handling operations can be performed, such as mixing, incubation, transportation, and storage [64].

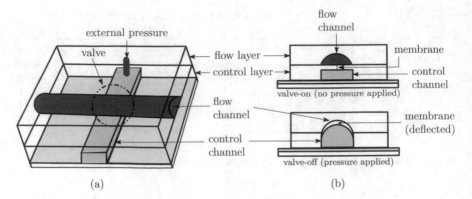

Fig. 1.3 Schematic of a two-layer microfluidic device: (**a**) top view and (**b**) cross-sectional view of valve states

1.4.3 Biochip Cyber-Physical System

A fully integrated biochip system consists of a controller, sensor feedback, and network connection [140], as shown in Fig. 1.1. A bioassay description (represented as a sequencing graph [36]) is synthesized to an actuation sequence that realizes the bioassay through various fluidic operations on the biochip. However, fluidic operations are susceptible to multiple manufacturing imperfections, which can lead to run-time faults. To detect such defects and for error recovery in the biochip operations [71], run-time monitoring through sensor feedback is required. CCD camera and/or capacitive sensors are used to monitor a droplet location and size on the DMFB [89]. CCD cameras are more popular due to their precision [89]. The image is cropped into sub-images to focus on an area-of-interest (e.g., an electrode). These sub-images are correlated with a template to monitor the droplet occupancy and size at the desired locations. The biochip can also be connected to a network for run-time monitoring of assay operations, result analysis, and software update.

In continuous-flow-based biochips (CFMBs), valves are controlled using a pneumatic pressure source connected through solenoid valves. The solenoid valves are electrically actuated to toggle between high pressure and low pressure. The implementation of a bioassay on a CFMB requires transforming each assay into a sequence of fluidic operations. These operations are then mapped to a sequence of electrical actuations. These actuations open (close) the solenoid valves, which in turn close (open) the microfluidic valves on the CFMB. This regulates the flow of fluid as required to realize fluidic operations. The pneumatic interface connects the CFMB valves to the external solenoid valves. The interface connection can be a direct one-to-one map or a control-logic-based mapping. In a one-to-one map, each CFMB valve is connected to an external solenoid valve, as shown in Example 1.

Example 1 Consider a sample preparation biochip shown in Fig. 1.4. The CFMB has a mixer and multiplexer that selects from two fluid inlets. It consists of eleven

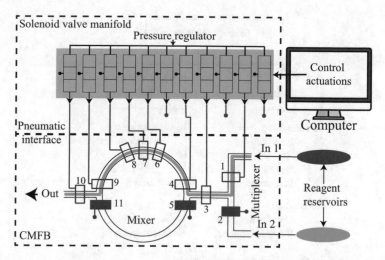

Fig. 1.4 A schematic of the one-to-one mapping between CFMB valves and external solenoid valves

valves. Each valve is connected to a solenoid valve array, i.e., eleven solenoid valves. These solenoid valves are part of a manifold(s) and can be individually actuated through a microcontroller to pressurize or depressurize the CFMB valves.

1.5 Related Work

Given the security-critical nature of the biochip application, the research community has focused its efforts on discovering its attack space and devising countermeasures. Figure 1.5 summarizes the footprint of the biochip security research so far.

1.5.1 Attacks

The earliest work on biochip security identified that biochips are susceptible to various attacks such as denial-of-service, contamination, and result manipulation [33]. These attacks can be launched in the following ways: (1) remote attackers can use malware to gain control of the network and to manipulate the control software or stored actuation sequence [55, 78, 117, 124]. (2) Proximity attackers can compromise the biochip by inducing faults in the controller or actuators using electrical probes or lasers [94, 119, 120]. As the biochip market expands, various design stages are expected to be outsourced, similar to integrated circuit design flow, to achieve faster design cycles. This leads to major security concerns of IP

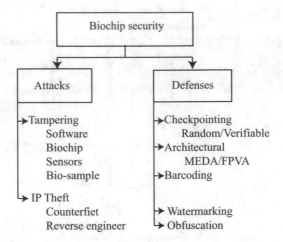

Fig. 1.5 Taxonomy of the microfluidic biochip security

theft and Trojan insertion. Trojans can be inserted into the CAD software or the biochip hardware. Threats to IP in the horizontal supply chain include overbuilding, reverse-engineering, and counterfeiting [31, 33, 96]. These attacks have been shown to be plausible on multiple biochip architectures [94, 97, 119, 120].

1.5.2 Defenses: Tampering

Software Checkers An actuation tampering can also be detected by crosschecking the actuation sequence before loading it to the DMFB. This can be done using a software checksum. However, software checkers cannot detect hardware fault injection attacks [32, 121], as software checkers are oblivious to changes down the control path. Other software-based defenses such as encryption techniques used for software integrity can be undone as the actuation sequence has to be decrypted before applying it to the DMFB. This leaves the decrypted actuations susceptible to tampering.

Remote Verification Recall that the DMFBs can be equipped with sensors that allow us to monitor its state. The sensor data of the biochip can be captured and stored, which can be validated later. This mechanism cannot support run-time error recovery as the sensor results are not processed for run-time fault detection. This makes the product susceptible to faults arising due to natural variation and hardware imperfections. In other words, run-time fault detection is indispensable in a biochip system [89] and remote verification does not make use of it.

Table 1.1 presents a comparison of defense strategies against biochip tampering. The earliest checkpoint defense uses CCD camera to monitor the DMFB at random

Table 1.1 Comparison of defense mechanisms

	Tampering attack detection		Error	Application scenario		
	Software [32]	Hardware [120]	recovery [89]	Lab	Online	Offline
Software checker [138]	✓	×	×	✓	✓	✓
Run-time checkpoint [121]	✓	✓	✓	✓	✓	✓
Remote verification [35]	✓	✓	×	✓	✓	×

Fig. 1.6 A DMFB cyberphysical system with multiple defenses. The software checker defense is in red, and the remote verification is in blue. The checkpoint-based defense is in green, which replaces the dotted portion of the control loop

time steps [123, 124]. The DMFB snapshots are processed to determine the run-time state of the biochip. By comparing the run-time state against the golden state over the entire execution cycles, the bioassay execution is validated. However, most biochip systems have minimal computing resources to minimize cost. Due to the time required for image capture and processing, continuous run-time monitoring of all DMFB cells is not possible. For example, the work in [121] shows that no more than 20 checkpoints can be examined in an execution cycle. This constraint was derived by considering an image pattern matching algorithm implementation on a mid-range ARM Cortex-M3 microcontroller. This constraint can be overcome by either algorithmic innovations that boost the probability of detection or enhanced biochip architecture for security (Fig. 1.6):

Algorithmic Innovations Heuristic defenses based on *checkpoints* are employed [87, 121]. Here, a spatial and temporal subset of the steps executed on the DMFB is sampled and used to compare against the golden specification. Algorithms based on randomized, weighted, and module-less choices have been proposed to derive the checkpoints [40, 87, 121]. These methods help in boosting the probability of attack detection.

Enhanced Architectures Another way of increasing the chances of attack detection is to use a platform with constrained capabilities. For this purpose, pin-constrained designs have been explored to maximize the probability of attack detection [122]. However, these measures are resource limited and do not guarantee a non-zero probability of attack evasion. A micro-electrode-dot-array (MEDA) platform offers a stronger proof of security with its integrated sensing [92, 97]. MEDA fabrication is costly due to the underlying CMOS layer; thereby, it could be cost prohibitive.

1.5.3 Defenses: IP Theft

Bioassay locking and watermarking are measures that protect the IP rights of a commercial product designer. Watermarking of bioassay was introduced to protect the IP rights of the developer [114]. Here, a secret message, which can be attributed to the IP owner, is embedded hierarchically into the biochip system. A watermark serves as proof-of-ownership in a court of law. A bioassay locking scheme was proposed to obfuscate the sequencing graph description of the bioassay [38]. Bioassay locking defends against an overproduction attack, such that an untrusted foundry cannot overproduce biochip hardware and sell it for profit. However, this technique does not prevent or resist reverse-engineering of a bioassay from the corresponding actuation sequence.

To prevent reverse-engineering of a bioassay implemented in a lab, a method for camouflaging the biochip layout by inserting extra valves and channels is reported in [43]. However, this approach fails as an attacker can reverse-engineer the IP by combining the actuation sequence and biochip layout. Another potential defense is to add extra actuations on idle valves to confuse the attacker. However, the attacker will be able to reverse-engineer the bioassay, albeit with the added extra operations. Careful insertion of dummy valves breaks the one-to-one mapping between the actuation sequence, biochip layout, and fluidic operations. In other words, the use of sieve valves (dummy) along with normal valves obfuscates the biochip layout and the actuation sequence. Without the knowledge of the type of the valve (normal/dummy), fluidic operations cannot be determined [93].

An orthogonal approach to our work can be to use 3D microfluidic design technique to obfuscate by distributing the design over multiple layers [95]. However, 3D fabrication is more time consuming and labor intensive, requiring multiple lithography steps and precision alignment [63].

Chapter 2
Threat Landscape

This chapter draws the threat landscape of the biochip systems, thus, provides the motivation and the scope of the book. Threats are described by modeling—*What* type of attacks are possible? *Where* in the system attacks are possible? *How* these attacks are executed? *Who* is responsible and *why*? These threats may lead to a loss of revenue or jeopardize the well-being of its users. More importantly, these erode trust between the stakeholders: designers, users, regulators, and investors. The objective of uncovering these attacks is to equip the stakeholders with necessary defense, thereby, boosting the overall trust in biochips.

Biochip threats can be classified into three main categories: online tampering, fabrication tampering, and reverse-engineering. Online tampering refers to either network-based attacks on biochip systems or physical device tampering. For example, a remote user can tamper with the controlling software or a malicious end-user can tamper with the biochip platform, the sensors, and the controller. Fabrication tampering refers to microfluidic Trojan insertion during the fabrication of a biochip. For example, a rogue actor in a fabrication unit can insert stealthy modification in the biochip—called Trojans—or a bio-sample vendor can provide tampered samples. Reverse-engineering refers to biochip IP theft. For example, stealing sensitive research data leads to loss of revenue and counterfeit products entering the market. In the subsequent sections, threat models of these attack types are described along with real-life case studies.

2.1 Online Tampering

Who presents risks and why? The attacker—who is in a remote location or near the biochip—could be a competitor seeking to bring disrepute to the biochip designer [131]. The proximity attacker can be an insider seeking to harm the end-user by manipulating the biochip results [110] or by denying service (DoS attack).

© The Author(s), under exclusive license to Springer Nature Switzerland AG 2022
S. Mohammed et al., *Security of Biochip Cyberphysical Systems*,
https://doi.org/10.1007/978-3-030-93274-9_2

How is the attack executed? The biochip CPS comprises controllers, software, network interface, sensors, and pneumatic actuators. The designer can source them from third-party vendors [31]. One can connect the biochip to a network for software updates and process the results online. Informed by this biochip supply chain, the attacker can launch an attack as follows:

1. Exploit the in-built hardware or software Trojan to access the biochip controller [121].
2. Use malware to gain control of the network and to manipulate the control software or stored actuation sequence [78].
3. Compromise the biochip by inducing faults in the controller or actuators using electrical probes or lasers [119].
4. Tamper the pressure inputs of the valves and pumps, causing undesired actuations [94].

What are the constraints on an attacker? The attacker manipulates the results of the biochip in a stealthy and untraceable way. To do this, the attacker has to evade detection by the sensors. The defender can monitor the biochip using a CCD camera and pressure sensors [88, 89]. **Who are the trusted actors?** The biochip designer is the defender. The designer trusts the biochip platform and the end-user.

Based on the above threat model, an attacker can maliciously modify the actuation sequence referred to as actuation tampering. This can be achieved through low-level changes to the stored actuation sequences in memory, or through modification of the control software. An attacker can aim for the following outcomes:

1. *Parameter Manipulation:* A critical step in bio-protocol development lies in the choice of parameters that optimize the assay. This requires systematic exploration of the design space to understand the interplay between parameters [19], which include mixing time, incubation time, mixing ratio, reagent volume, and concentration. The developer, after many trials, determines the parameter values. Any tampering of the parameter values during the execution of the bio-protocol on a biochip can invalidate the outcome.
2. *Contamination:* A bioassay involves the transportation of different fluids along predetermined paths in the biochip. Deviations from these paths can cause contamination of the fluids, yielding wrong results.
3. *Miscalibration* In bioassays such as diagnosis, a calibration curve is used to interpolate the results of the sample-under-test. An attacker can tamper with the curve to enforce misinterpretation of results.

2.1.1 Case Study: Immunoassay

The immunoassay protocol relies on the antigen–antibody interaction. Antibodies are proteins that are generated by the immunocells of our body in response to the invasion of an antigen. High specificity in the binding of an antigen with an

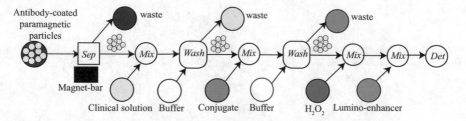

Fig. 2.1 A DMFB implementation of the immunoassay. *Mix*, *Wash*, and *Det* represent mix-incubate, wash, and detection operations, respectively. *Sep* is the operation for separating particles from diluent by magnets

antibody enables quantitative measurement of the target antigen. A bench-top non-competitive immunoassay protocol performs a sequence of incubation and washing steps. The DMFB implementation of immunoassay protocol differs from its bench-top implementation. For example, the DMFB completes the washing operation by using antibody-immobilized magnetic beads. We illustrate a paramagnetic bead-based DMFB adaptation of the immunoassay [99].

Figure 2.1 shows the sequencing graph of a microparticle-based immunoassay. First, a droplet encapsulating the antibody-attached paramagnetic particles is dispensed. The DMFB separates the particles from the diluent using a magnet. Next, a clinical solution that may contain the antigen/protein of interest is dispensed and mixed with the particles and incubated. The particles are washed in the wash buffer to separate any unbounded protein/antigen. Next, a droplet of a conjugate solution containing enzyme-linked antibody is dispensed and mixed with the washed particles. The particles are then washed and suspended in the wash buffer and queued for analysis. The DMFB separates the particles from the wash buffer and mixes it with a droplet of H_2O_2 and incubates. This droplet is mixed with a droplet of a luminol-enhancer solution and incubated. Finally, the DMFB records the chemiluminescence for quantitative measurement using a calibration curve. The assay completes within a few hours.

2.1.1.1 Parameter Tampering

In the immunoassay protocol described in Fig. 5.6, the fluid is incubated after mixing with a lumino-ehancer. Increasing the incubation time by 2 min causes the chemiluminescent signal to drop by up to 45% [46]. This way, the attacker can alter the inference drawn by the bio-protocol. Since the results are not out-of-range (unlike in a DoS attack), the user does not suspect foul play, i.e., the attack is stealthy.

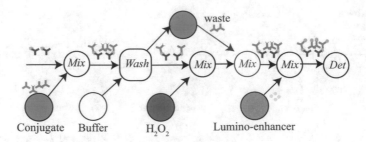

Fig. 2.2 The contamination attack in immunoassay through extra mix operation Mix shown in red

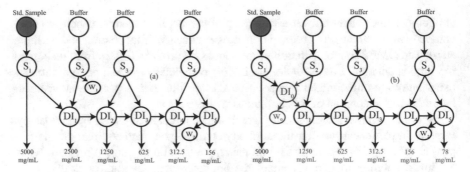

Fig. 2.3 Miscalibration attack. (**a**) Original dilution graph. (**b**) Miscalibrated dilution graph. Here, S_i represents a split operation, and DI_i represents a dilution operation. In the miscalibrated scenario, the product of split operation S_1 is subjected to an extra dilution operation DI_0. This reduces the concentration of the product of subsequent dilution operations DI_1, DI_2, etc

2.1.1.2 Contamination

In DMFB-based immunoassay described in Figure 5.6 involves the washing of unbounded conjugate antibodies, as shown by the snippet in Fig. 2.2a. This forms the waste droplet shown in green in Fig. 2.2. If this waste droplet is routed to be mixed with the washed solution, then the final output is contaminated, as shown in Fig. 2.2b. Even if the sample did not contain the targeted antibody, the color in the final output could be inferred as the presence of the antibody. This can lead to false-positive detection of the antibody.

2.1.1.3 Miscalibration

In the immunoassays, a calibration curve is an output of a regression, where the detection of different known protein concentrations is performed. In these experiments, various concentrations of serially diluted standard solutions are used, as shown in Fig. 2.3a. In each experiment, the optical intensity of the fluorescence signal is used against the corresponding protein concentration to generate the calibration curve. This curve is used to estimate the concentration of the protein in a sample-under-test. An attacker can manipulate the dilution graph, as shown in Fig. 2.3b. This changes the gradient slopes of the calibration curve and causes misinterpretation of the results.

2.2 Fabrication Tampering

The following items describe the threat model: **Where?**—Consider a biochip developer who outsources the fabrication of the biochip to an untrusted foundry. **Who and Why?**—A rogue element in the foundry tampers with the biochip design to insert a Trojan. The objective of the attacker could be to bring disrepute to the biochip developer [110]. **How?**—The attacker has access to the design files through which he/she could identify the biochip components. The Trojan trigger can occur naturally in a bioassay implementation, or the control software can be modified remotely to trigger the Trojan, as shown in Fig. 2.4. **Limitations**—However, the attacker has a constraint that the biochip needs to pass the post-manufacturing fault testing. We assume that a trusted actor performs the test of the biochip. **What?**—The Trojan attack can be used to perform the attacks described in the previous section, namely parameter manipulation, contamination, and miscalibration.

2.2.1 Case Study: Microfluidic Trojan

A Trojan is defined as a malicious modification of the design. It can be divided into two components: trigger and payload. The basic idea of a microfluidic Trojan is to use a multi-height (thinner) valve as a Trojan payload, which begins to leak when the valve pressure drops due to activity on the other valves. We explain the design through the following two concepts:

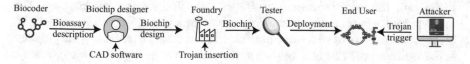

Fig. 2.4 Biochip design supply chain and threat model for the microfluidic Trojan-based attack

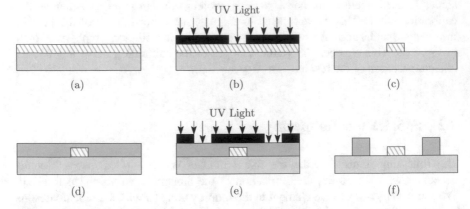

Fig. 2.5 Fabrication of multi-height structure: a cross-sectional view of (**a**) spin coating with photoresist, (**b**) UV exposure, (**c**) washing of uncured photoresist, (**d**) spin coating with photoresist, (**e**) UV exposure, and (**f**) washing

2.2.1.1 Design of a Multi-Height Valve

An attacker can maliciously lower the height of the control channel, as shown in Fig. 2.5. This results in a thicker membrane that requires a higher pressure to operate (close/open) compared to the normal membrane. When it is operated at a lower pressure, the valve does not close/open completely [72, 79]. For example, the work in [79] shows that a 34 μm membrane valve requires a minimum pressure of 12 psi to operate, whereas a 28 μm membrane requires a minimum pressure of 8 psi to operate. Such valve misbehavior can hamper biochip functioning. The proposed microfluidic Trojan payload is designed based on this phenomenon.

2.2.1.2 Draining of Valve Pressure

The proposed microfluidic Trojan is triggered in two ways: operation based or time based. The valve state can be closed (opened) when the pressure at the valve is high (low). The valve pressure is controlled through the pneumatic interface. The pressure source is connected to multiple valves through a solenoid valve manifold. If the valves associated with a manifold close/open at a high frequency, the pressure at the source drops. This is analogous to the draining of voltage at a battery when a large current is drawn from it. This is referred to as an *operation-based trigger*.

In the case of the control logic interface, the pressure source is connected to the valve selected through the control logic. Recall that such valves latch their states and need a periodic refresh. However, as the membrane height increases, it requires a more frequent refresh, else the valve pressure drops [67]. Suppose the normal valves need to be refreshed at rate T to retain their valve state. Due to its thicker membrane, the payload valve will need a frequent refresh rate τ ($\tau < T$). If the refresh rate (t) is in the range $T > t > \tau$, the payload valve can be triggered after τ time. This is referred to as a *time-based trigger*.

2.2.1.3 Attack Model

Payload A rogue element in the foundry inserts the Trojan by identifying one or more valves suitable for the attack objectives. The attacker increases the valve (Trojan-payload) membrane height to the maximum extent that allows the device to pass the post-manufacturing test. When the biochip is deployed in operation, the Trojan can be triggered by either time-based or operation-based draining of the pressure source. The microfluidic Trojan attack is stealthy because there is no visible change in the payload valve, so the online sensor-based monitoring cannot detect it [113].

Trigger The trigger conditions can occur naturally in a bioassay that has been synthesized for high throughput (tighter refresh cycles) and lower overhead (multiple valves share a pressure source). The Trojan gets activated for a very short time, causing an intermittent fault. Current testing and functional methods are not designed for detecting intermittent faults [72, 88]. Therefore, it is unlikely that the designer considers the change in timing and pressure behavior of the valve.

2.3 Reverse-Engineering

Continuous flow-based microfluidic biochips (CFMBs) have evolved rapidly in the last decades [90, 91]. The CFMBs allow automated control of fluid flow in a network of microchannels by suitable actuation of pressure-driven microvalves [91]. Bioassay implementation on a CFMB can be reverse-engineered using biochip images and actuation sequence [43]. The following items describe the threat model: **Where?**—Consider a bioassay developer who invests heavily in bioassay IP development and uses microfluidic platforms to conduct large-scale experiments involving the bioassay [90]. **Who?**—A competitor is motivated to steal the IP from the developer without incurring any cost of development. **What?**—To RE the bioassay, the attacker accesses the actuation sequence and the video or snapshots of the biochip captured by the camera sensor. **How?**—An attacker can get the "what" through a network attack. The biochip controller is connected to the network for round-the-clock online monitoring and control [17]. A remote attacker can gain administrator credentials using a social engineering attack or malware [78, 103]. Alternately, an attacker can collude with rogue insiders to capture the video of bioassay execution. Image analysis-based RE can recover the actuation sequence and the bioassay [43]. **Limitations**—The attacker can differentiate between the actuated valve and deactivated valve due to the visual difference. However, the attacker does not know which one is a normal valve or which one is a sieve valve as the top view of both is identical (Fig. 1.3). Further, the attacker cannot obverse the flow of fluid through the channels as it does not produce any visual difference. The attacker does not have access to the CFMB.

2.3.1 Case Study: Bioassay Theft

We demonstrate IP piracy through a bioassay implementation on a CFMB (Fig. 2.6). The platform consists of a multiplexer that selects from two input reagents R_1 and R_2 and uses a rotary mixer to mix them in the desired ratio [130]. Fluidic operations corresponding to a bioassay are mapped to a sequence of actuation steps for controlling the valve state. Let us illustrate the bioassay execution on the CFMB in which all valves are initially closed. The first set of actuations fills R_1 in the upper half of the mixer (Fig. 2.6a). Next, R_2 fills the lower half of the mixer (Fig. 2.6b). The valves 6, 7, 8 are activated in a sequence to form a peristaltic pump that circulates the fluid in the rotary mixer, producing a mixture of R_1 and R_2 in a 1:1 ratio (Fig. 2.6c). Next, the lower half of the mixer is replaced with R_1 (Fig. 2.6d), and the peristaltic pump is activated (Fig. 2.6e). The resulting fluid contains R_1 and R_2 in a 3:1 ratio (Fig. 2.6f).

Figure 2.6 shows the straightforward one-to-one mapping between the actuation sequence, biochip snapshots, and fluidic operations. It can be inferred from the biochip snapshots that the bioassay mixes two input fluids in a 3:1 ratio [43]. The corresponding sequencing graph (IP) is shown in Fig. 2.6e. The mixing time can also be determined from the actuations. This example demonstrates the ease with which the bioassay description and its parameters can be reverse-engineered [43]. To thwart the RE of bioassays, we need to obfuscate the one-to-one mapping between the actuation sequence, biochip snapshots, and fluidic operations.

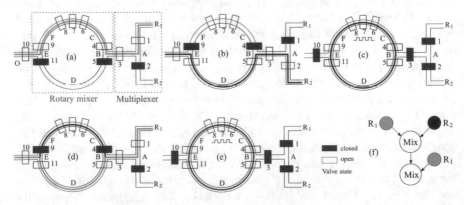

Fig. 2.6 A bioassay implementation: (**a**) Push reagent R_1 into the upper half of the mixer. (**b**) Push reagent R_2 into the lower half of the mixer. (**c**) Mix. (**d**) Push R_1 into the lower half of the mixer. (**e**) Mix. (**f**) The sequencing graph is inferred from the images

Chapter 3
Architecture for Security

Biochip system design flow needs to be security aware to address various threats. This can be achieved by using a biochip architecture that is more amenable to the system's security needs. Biochip architectures vary in features such as size, degree of programmability/controllability, sensor type/granularity. Each of these features has implications to the overall system security. Therefore, security assessment of various biochip architectures is required, so that an appropriate biochip is used.

3.1 Security Metric

Recall that a biochip CPS consists of a controller, sensor feedback, and network connection [140]. Fluidic operations implemented on a biochip are susceptible to multiple manufacturing imperfections, which leads to uncertainty. To detect such defects and for error recovery in the biochip operations [71], run-time monitoring through sensor feedback is required. For example, CCD camera and/or capacitive sensors are used to monitor a droplet location and size on the DMFB [89]. CCD cameras are more popular due to their precision [89]. The image is cropped into sub-images to focus on an area-of-interest (e.g., an electrode). These sub-images are correlated with a template to monitor the droplet occupancy and size at the desired locations.

The online monitoring infrastructure can also be used for detecting tampering attacks by comparing the run-time droplet locations against a golden reference. The biochip designer generates the golden droplet map from the golden actuation sequence and stores it in a security co-processor, which is physically separated from

---

Based on "Towards Secure Checkpointing for Micro-Electrode-Dot-Array Biochips," TCAD, 2020 [116].

S. Mohammed et al., *Security of Biochip Cyberphysical Systems*,
https://doi.org/10.1007/978-3-030-93274-9_3

Fig. 3.1 Probability of
evasion (P_E) decreases with
an increase in the deviant
cells (ΔN) for a given sensor
coverage ($\frac{k}{T}$)

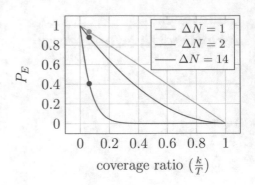

the biochip controller. The security co-processor is not connected to the network. This prevents the attacker from compromising both the bioassay execution and the defense [121].

The granularity and the extent of monitoring depends on the sensor availability and controller's computational capacity. This limitation defines the security metric—*the probability of attack evasion*. Consider a DMFB with a total of "T" cells. Let sensing capability allow for "k" cells to be monitored in a given time-step. The probability that a random cell is being monitored is given by $\frac{k}{T}$. The probability that a random cell is overlooked, i.e., the probability of evasion P_e is given by the $\left(1 - \frac{k}{T}\right)$. The attacker's objective is to escape the detection scheme. An ideal attack is one where there is minimal change in the droplet locations. If there are ΔN deviant cells in a time-step, the probability that all of them go undetected is given as follows:

$$P_E = \left(1 - \frac{k}{T}\right)^{\Delta N} \tag{3.1}$$

Figure 3.1 shows that the online monitoring approach provides some assurance of integrity; however, the capabilities of DMFB hardware limit the probability of successful attack detection. The achievable probability of detecting an attack can be as low as 50% on a realistic embedded DMFB controller. This is not satisfactory and not likely to inspire confidence in users of these DMFB systems. Informed by this metric, the rest of the chapter presents a security assessment of next-generation DMFB called micro-electrode-dot-array (MEDA). The security assessment template can be captured in the below steps:

1. Identify the unique features and operations that can be performed on biochip architecture.
2. Expand the attack space that is enabled by the unique features.
3. Calculate security metrics that capture the attack stealthiness and effectiveness of the possible defense.

4. Expand defenses that utilize available resources to defend against the expanded attacks.

This template can be applied to any arbitrary biochip architecture [94, 119, 120].

3.2 Micro-Electrode-Dot-Array

Traditional DMFBs suffer from several disadvantages that limit their scalability and reconfigurability: (1) droplet size is constrained by the electrode size, (2) droplet volume control is limited, and (3) sensors must be integrated post-fabrication [82]. Micro-electrode-dot-array (MEDA) biochip overcomes these drawbacks.

A MEDA biochip platform consists of the following components: (1) a two-dimensional array of identical microelectrode cells (MCs) and (2) a biochip controller, which consists of a chip layout map, droplet location map, and the fluidic operation manager, as shown in Fig. 3.2a. Each MC includes a high voltage-driven microelectrode, an actuation circuit, and a sensing circuit for real-time sensing of the droplet under the microelectrode, as shown in Fig. 3.2c. Depending on the application, MCs are grouped to form a virtual chip layout that contains reservoirs, mixers, and fluidic paths. The chip layout map stores this configuration data. The droplet location map stores the real-time locations of the droplets provided by the sensors in the MCs. The fluidic operation manager converts a fluidic instruction to an MC actuation pattern, which is then shifted into the MCs. The MCs loaded with logic "1" are actuated by connecting high voltage. After the actuation, the sensor is enabled, and the data is shifted out, creating the droplet location map, as shown in Fig. 3.3a. The load, execute, and sense steps form an actuation cycle.

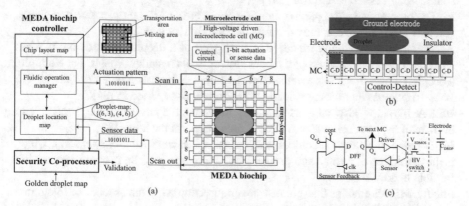

Fig. 3.2 MEDA cyberphysical system. (**a**) MEDA CPS includes the MEDA biochip, the controller, and the security co-processor. (**b**) Side view of the MEDA biochip. (**c**) A circuit schematic of the sensor and control module comprising a micro-electrode cell (MC). Q_n denotes the nth cell in the scan chain, receiving an input from the $(n-1)$th cell denoted by Q_{n-1}

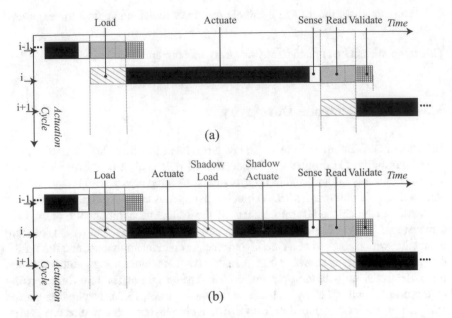

Fig. 3.3 Original and shadow-attack-modified actuation timeline. (**a**) The baseline MEDA actuation cycle. (**b**) The modified MEDA actuation cycle with a "shadow operation" embedded in the time slack (in red)

3.2.1 Unique Features of MEDA

MEDA biochips have a large number of microelectrodes, which are 10–20 times smaller than DMFB electrodes [137].

MEDA biochip supports precise and flexible control of the droplet size and shape, whereas DMFB supports only droplets of a single size and shape.

According to the velocity model [82], the speed of a droplet depends on its mass and size. Hence, a larger droplet moves slower than a smaller droplet. For example, a droplet of size 5×5 has a velocity of 1.1 mm/s. Whereas, a larger droplet of size 16×16 size has an average velocity of 0.6 mm/s, as reported in [81].

MEDA biochips also support diagonal movement apart from the vertical and horizontal movements, whereas DMFB only supports the latter [82].

MEDA also introduces specific fluidic operations such as *aliquot operation*, which is the extraction of a small sub-droplet from a large droplet. Such operations are not feasible in a DMFB.

Finally, MEDA enables fine-grained sensing in contrast to the absence of integrated sensing in DMFB [141].

Table 3.1 Digital microfluidic attack space

Attack space							
Types	Macro [117]			Micro [117]	Shadow [111]		
Sub-types	Add	Modify	Skip	Aliquot	Swap	Merge	Aliquot	Split
Effects	High result deviations			Minute result deviations	Variable result deviations		
Corresponding defenses							
DMFB	Probable security by randomized CPs [121]			Not applicable	Not applicable		
MEDA	Provable security by graph reconstruction [85]			Variability-aware droplet map comparison [117]	Shadow-attack-aware extra checkpoint insertion		

3.3   MEDA Attack Space

Based on the threat model described in Chap. 2, we sketch the attack space for the MEDA biochip. We define two new classes of actuation tampering attacks for MEDA biochip: granular and shadow attacks. The attack space and their corresponding state-of-the-art defenses are shown in Table 3.1. These are briefly described as follows:

3.3.1   Granular Attacks

MEDA biochips offer fine-grained droplet control. The attacker can leverage this to launch attacks of varying granularity.

Macro-droplet attacks are malicious modifications to the actuation sequence that operate on whole droplets. This can be done by performing one of the following actions: *add*, *modify*, and *skip* operation(s) specified in the assay. This results in a major deviation in the droplet map and is similar to the DMFB threat models discussed in earlier work [32, 121], albeit for an arbitrarily sized droplet.

Example 2 Consider in vitro glucose measurement assays to determine blood glucose levels. In this assay, a glucose calibration curve is constructed, which plots the rate of reaction for various concentrations of serially diluted standard glucose solution. This curve is used to interpolate the concentration of the glucose sample-under-test. An example of a macro-droplet attack dilutes sample droplets with a waste droplet during calibration, which is similar to the attack shown in Fig. 2.3 in Chap. 2. The glucose concentration readings can deviate by 50–100% from the golden value [32].

Micro-droplet attacks extract a small droplet from a larger droplet using an aliquot operation [141]. These attacks are difficult to detect since droplets naturally lose volume through cutting or evaporation during movement [133]. Furthermore,

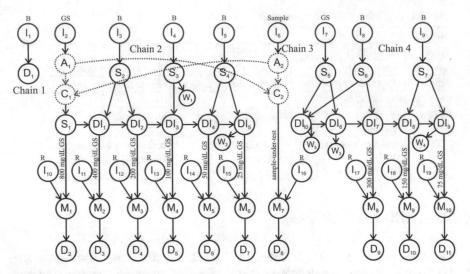

Fig. 3.4 A micro-droplet attack on the glucose assay. The attack manipulates the concentration of droplets GS and sample through aliquot operations A_1, A_2 on these droplets, respectively. The aliquot droplet corresponding to GS is mixed with the sample droplet through a contamination operation C_2. Simultaneously, the aliquot droplet corresponding to the sample droplet is mixed with the GS droplet through a contamination operation C_1. The malicious modification of the glucose assay is highlighted by the red dotted lines

the movement of micro-droplets is fast compared to large droplets. Therefore, micro-droplet attacks are stealthy and easy to launch.

Example 3 Consider the glucose assay similar to the previous example. The attacker alters the actuation sequence to perform an aliquot operation on GS and merges it with the sample droplet S. Simultaneously, the attacker performs an aliquot operation on the sample S and merges it with the GS droplet, as shown in Fig. 3.4. Since the aliquot droplets are smaller, they introduce smaller degree of contamination compared to contamination by merge–split operations. The micro-droplet attack leads to a very subtle variation in the bioassay result compared to that of the macro-droplet attack [32]. Further, the attacker could perform multiple aliquot operations to manipulate results with fine granularity. Therefore, the attacker can achieve a range of targeted deviations through micro-droplet attacks. Such stealthy attacks would lead to poor patient outcomes, especially considering that 60–70% of medical care decisions are dictated by test results [61].

3.3.2   Shadow Attacks

As the MEDA biochip supports droplets of varying sizes and velocities, attacks can exploit slack in faster droplets to execute extra operations. The state of the MEDA biochip at time t consists of the droplet map and a unique identifier associated with each droplet on the MEDA biochip at that time instant. The state at time t, $S^t = \langle L_S^t, I_S^t, \delta_S^t \rangle$, where L_S^t, I_S^t, and δ_S^t is the droplet map, and the set of all droplet identifiers at time t, and a function that maps the droplet identifier to the location. At time t, the *shadow state* $S^t = \langle L_S^t, I_S^t, \delta_S^t \rangle$ of the golden state $G^t = \langle L_G^t, I_G^t, \delta_G^t \rangle$, if $L_S^t = L_G^t$, $|I_S^t| = |I_G^t|$, and $\delta_S^t \neq \delta_G^t$. Hence, the shadow state has a similar droplet map to the golden state. An assay is a sequence of state transitions wherein a state S^t transitions to state S^{t+1} on a fluidic operation F^t. That is, $S^t \xrightarrow{F^t} S^{t+1}$. If the state S^t transitions to state $\tilde{S}^{t+1}$ with the same droplet map as S^{t+1} on a different set of fluidic operations $\tilde{F}^t$, then $\tilde{F}^t$ is a shadow operation. $S^t \xrightarrow{\tilde{F}^t} \tilde{S}^{t+1}$, where $L_S^{t+1} = L_{\tilde{S}}^{t+1}$.

Example 4   Consider a 10×8 MEDA biochip in which three droplets (a large 3×4 size droplet and two smaller 2×2 size droplets) are moved in an actuation cycle. The MEDA biochip has an electrode pitch size of $50\,\mu m$, and the spacing between the plates is $50\,\mu m$. The three droplets can be the same reagent of varying concentrations in a sample preparation protocol, which have similar viscosity and interfacial tension. Figure 3.5a, c shows the initial and final states of the golden actuation cycle. The larger droplet moves slowly relative to the smaller ones because of the greater resistance and viscous drag that it runs into. Let the smaller droplet (2×2) move at an average speed of 1.3 mm/s, and the larger droplet has an average speed of 1 mm/s. Figure 3.5b presents the transitional state in which the two smaller droplets reach their destinations. Without loss of generality, we estimate that the smaller droplets move two times faster than the larger droplet. Hence, the smaller droplets travel two steps in the time it takes for the larger one to finish one step. This sets up a timing slack for the smaller droplets in the actuation cycle.

An attacker can manipulate the timing slack to carry out shadow transport operations on the smaller droplets. The attacker can uphold the golden droplet map at the conclusion of the actuation cycle. Figure 3.5e–f shows one possible malicious droplet movement that interchanges the destinations of the two smaller droplets while keeping the droplet map at the end of the actuation cycle (parts c and f in Fig. 3.5 have identical maps).

Let L_A^t (L_B^t) and L_A^{t+1} (L_B^{t+1}) be the locations of two equal-sized droplets A (B) at cycle t and $t + 1$, respectively. The average speed of the droplet A (B) is given by v^A (v^B) and t_{cycle} is the each cycle duration. Depending on which shadow operation is performed in the available slack, different shadow attacks can be launched:

1. *Swap:* where droplets interchange their respective locations, as shown in Example 4 and Fig. 3.5d–f.

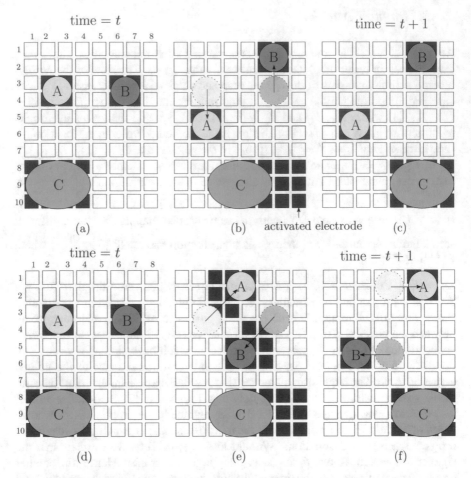

Fig. 3.5 Droplet transport on a MEDA biochip: (**a**) the initial state, (**b**) the intermediate state, and (**c**) the final state of an actuation cycle. A shadow operation during the droplet transport: (**d**) the initial state, (**e**) the intermediate state, and (**f**) the shadow state at the end of an actuation cycle

$$\left(|L_A^t - L_B^{t+1}| < v^A \cdot t_{cycle}\right) \quad \wedge \quad \left(|L_A^{t+1} - L_B^t| < v^B \cdot t_{cycle}\right) \tag{3.2}$$

2. *Split–merge:* where two droplets are split in t_{split} time and the resulting child droplets of one fluid are merged with the child droplets of the other fluid. This results in two droplets that are a mix of two child droplets, as shown in Fig. 3.6d–f.

$$\left(|L_{A_1}^* - L_B^{t+1}| < v^{A_1} \cdot (t_{cycle} - t_{split})\right) \wedge \left(|L_{B_2}^* - L_A^{t+1}| < v^{B_2} \cdot (t_{cycle} - t_{split})\right) \tag{3.3}$$

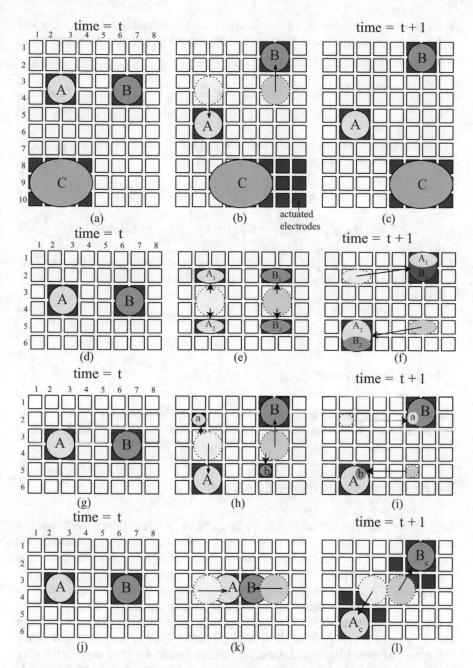

Fig. 3.6 Illustration of shadow operations. (**a**)–(**c**): Snapshots of the golden execution. (**d**)–(**f**): A split–merge splits droplets A and B and then merges child droplets. (**g**)–(**i**): An aliquot-merge extracts aliquot droplets from A and B and merges them with the other droplets. (**j**)–(**l**): A merge–split contaminates droplets A and B by merging them and splitting

3. *Aliquot-merge:* where an aliquot droplet (a, b) is extracted from two droplets (A, B) in $t_{aliquot}$ time, and these aliquot droplets are mixed with the other droplet (B, A). This is shown in Fig. 3.6g–i.

$$\left(|L_a^* - L_B^{t+1}| < v^a \cdot (t_{cycle} - t_{aliquot})\right) \wedge \left(|L_b^* - L_A^{t+1}| < v^b \cdot (t_{cycle} - t_{aliquot})\right)$$
$$(3.4)$$

4. *Merge–split:* where two droplets are merged (mid) and then split, resulting in contaminated droplets, as shown in Fig. 3.6j–l.

$$\left(|L_A^t - L_{mid}^*| + |L_{mid}^* - L_A^{t+1}| < v^A \cdot (t_{cycle} - t_{split})\right) \quad \wedge$$

$$\left(|L_B^t - L_{mid}^*| + |L_{mid}^* - L_B^{t+1}| < v^B \cdot (t_{cycle} - t_{split})\right)$$
$$(3.5)$$

5. *I/O-swap:* where a droplet can be swapped with a different droplet from an inlet I, and the swapped droplet pushed out to an outlet O.

$$\left(|L_I - L_A^{t+1}| < v^I \cdot t_{cycle}\right) \wedge \left(|L_A^t - L_O| < v^A \cdot t_{cycle}\right)$$
$$(3.6)$$

3.4   MEDA Defense

Having described the various types of attacks on MEDA biochip, we now describe an updated defense mechanism that inserts checkpoints to mitigate these attacks.

3.4.1   *Micro-Attack Aware Checkpoint*

An effective way to detect attacks is to compare the real-time sensed biochip droplet map against the golden droplet map. Deviations from the golden droplet map are flagged as attacks. Recall that natural variation in the droplet size is expected due to cutting or evaporation during movement or due to the faults in the biochip [83, 133]. Because of this uncertainty, a micro-electrode-to-micro-electrode comparison between the real-time sensed droplet map and the golden droplet map is not practical for validation. To overcome this, we divide the golden droplet map into three regions: (1) occupied electrodes, (2) empty electrodes, and (3) guard-band electrodes, as shown in Fig. 3.7. Occupied electrodes are the ideal expected droplet location, guard-band electrodes stretch this ideal location to cover the possible natural variations, and empty electrodes form the dead space on the biochip where droplets are not expected. To capture this logically, we store the golden droplet map

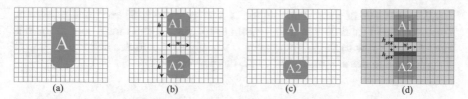

Fig. 3.7 Droplet map on MEDA biochip. (**a**) A droplet on a MEDA biochip before the split operation. (**b**) The droplet is split uniformly into two equal halves. (**c**) The droplet is split into unequal halves. (**d**) Droplet map that covers the ideal droplet location and the expected variations. The yellow grid elements are the empty electrodes, the green grid elements are the occupied electrodes, and the gray grid elements are the guard-band electrodes

with two bits for each micro-electrode. Occupied electrodes are denoted by "01," empty electrodes are denoted by "00," and a "1x" denotes a guard-band electrode.

3.4.1.1 Golden Droplet Map Generation

The golden droplet map is derived from the actuation sequence. In an ideal biochip, there is a one-to-one correspondence between the actuation sequence and the physical droplets. However, due to natural variations and errors, the true droplets will deviate from the actuation sequence [140]. The golden droplet map must accurately capture these variations or risk having the defense system raise false positives and negatives. Therefore, the golden droplet map will be a one-to-one copy of the actuation sequence, except for the guard-band region.

Assuming a droplet is moving along the y axis, the dimensions of the guard-band region can be calculated directly from the intrinsic error limit E_{lim} as follows:

$$h_{gb} = \lceil E_{lim} \cdot h \rceil$$
$$w_{gb} = w$$

(3.7)

where h and w are the ideal droplet height and width (Fig. 3.7b), h_{gb} is the derived guard-band height, and w_{gb} is the derived guard-band width (Fig. 3.7d). This guard band is a rectangular region that can be imagined as a "trail" behind the ideal droplet location. That is, the width is not prone to variation since it is orthogonal to the droplet movement, while the height may vary by as much as E_{lim}. The ceiling function is used to round up the height h_{gb} as we can only have an integer number of electrodes. The calculation for movement along the x-axis is performed analogously.

Example 5 In Fig. 3.7, droplet A of size 5×10 is split into two droplets $A1$ and $A2$. The ideal size of $A1$ and $A2$ is 5×5, which corresponds to the occupied electrodes of the golden droplet map, as shown in Fig. 3.7d. The size of the ideal droplet is $w = 5$ and height $h = 5$ (Fig. 3.7b). We consider $E_{lim} = 8\%$, as reported in [140]. From

Eq. (3.7), the size of guard-band region is $h_{g}b = \lceil 0.08 \times 5 \rceil = 1$ and $w_{gb} = 5$. The direction of fluid flow is from bottom to top for $A1$ and top to bottom for $A2$. Therefore, the row to the bottom of the $A1$ occupied electrodes and top of the $A2$ occupied electrodes are marked as guard-band electrodes.

3.4.2 Shadow Attack Aware Checkpoints

Recommended defense goes through the shadow operation possibilities in each actuation cycle and shrink the duration of those actuation cycles susceptible to shadow attacks. For each actuation cycle in the golden actuation sequence, we establish the number of droplets and their sizes, which determines their velocity. We use these parameters to check the condition of each shadow operation between all pairs of droplets appearing in the actuation cycle. Algorithm 1 outlines the defense.

A bioassay is executed on the MEDA platform by synthesizing the assay specification into a sequence of actuation cycles [81]. The actuation sequence has clearly delineated actuation cycle boundaries. A security co-processor checks the

---

Algorithm 1 Defend against shadow attacks with extra checkpoints

---

Input: Actuation sequence
Output: Shadow attack resistant actuation sequence
for *each actuation cycle in the actuation sequence* **do**
    t_{cycle} = time of the current actuation cycle $t = t_{cycle}$
    for *each droplet pairs A, B*
        in the current actuation cycle **do**
            `// Swap attack -- ` Eq. (3.2)
            if *Expression* (3.2) *is true for A and B* **then**
                | Calculate min cycle time to avoid swap. $t = \min\{t, t_{min}\}$
            `// Split-merge attack -- ` Eq. (3.3)
            if *Expression* (3.3) *is true for A and B after splitting* **then**
                `// Any one or both A and B can be split`
                Calculate min cycle time to avoid split–merge. $t = \min\{t, t_{min}\}$
            `// Aliquot-merge attack -- ` Eq. (3.4)
            if *Expression* (3.4) *is true for A and B* **then**
                | Calculate min cycle time to avoid aliquot-merge. $t = \min\{t, t_{min}\}$
            `// Merge-split attack -- ` Eq. (3.5)
             if *Expression* (3.5) *is true for A and B* **then**
                | Calculate min cycle time to avoid merge–split. $t = \min\{t, t_{min}\}$
    end
    for *each pair A, Inlet I, Outlet O*
        in the current actuation cycle **do**
            `// IO-swap attack -- ` Eq. (3.6)
            if *Expression* (3.6) *is true for A, I, O* **then**
                | Calculate min cycle time to avoid I/O-swap. $t = \min\{t, t_{min}\}$
    end
    if $t < t_{cycle}$ **then**
        | Split current actuation cycle into two cycles of lengths t and $(t_{cycle} - t)$, respectively.
end

---

golden map against the droplet map at the end of each actuation cycle. Synthesis algorithms for the MEDA biochip are oblivious to shadow operations [41, 76, 81]. Hence, the shadow operations can subvert the bioassay. The defense increases the number of actuation cycles in the golden actuation sequence, which increases the checkpoints and the space needed to store the golden maps.

Pruning of Checkpoint We next minimize the number of checkpoints by dropping those time-steps that are not susceptible to any attacks. Consider a case where droplet occupancy is sparse in a given time-step t. Let the time interval between current time-step ($t_{current}$) and previous checkpoint time-step (t_{prev_check}) be $t_{current} - t_{prev\_check}$. If in this time interval no attack is possible, i.e., the conditions (3.2)–(3.6) are not satisfied for $t_{cycle} = t_{current} - t_{prev\_check}$. Then, the time-step $t_{current}$ can be safely dropped from the checkpoint list. We use this observation to minimize the number of checkpoints. We initialize the checkpoint list with all time-steps, considered as baseline. We then iterate over each assay time-step $t = 1, 2, \cdots, T$. Time-step $t_{current}$ is dropped from the list if it is safe, else we update the latest checkpoint time-step as $t_{prev\_check} = t_{current}$ and move to the next time-step $t_{current} + 1$.

3.5 Experimental Results

3.5.1 Micro-Attack Aware Checkpoints

A bioassay execution is simulated on a DMFB biochip. The results of this simulation are used to project the implementation cost for a MEDA biochip. The assay was simulated for a 17×31 DMFB biochip using the open-source UCR Static Simulator [65, 66]. A 100 Hz clock was used to actuate the electrodes. The aliquot operation in the MEDA biochip was mimicked by a split operation on the DMFB. The aliquot operation takes 3–4 cycles, whereas a split operation takes one cycle. Since the assay execution takes a total of 1839 cycles, mimicking the aliquot operation by a split operation causes only a negligible error in the total execution time. The execution time for the golden assay and the assay with the micro-droplet attack is 18.39 s and 18.22 s, respectively. This small-time difference in execution time is consistent with our earlier assertion that the micro-droplet attack is stealthy.

To reflect the high density of micro-electrodes in a MEDA biochip, each DMFB electrode is divided into 4×4 micro-electrodes [76, 83]. Therefore, droplets occupying a single electrode in the DMFB now occupies 4×4 microelectrodes in a MEDA biochip of $(17 \times 4) \times (31 \times 4) = 68 \times 124$ size. Although this model does not capture the performance boost that MEDA offers through diagonal droplet movement, it is sufficient to study the cost of using fine-grained sensing for validation. The golden droplet map requires storage space of $68 \times 124 \times 2$ bits for each cycle. The total memory required to store the golden droplet map of the glucose measurement assay over 1839 cycles is $68 \times 124 \times 2 \times 1839 = 3.877$ M.

3.5.2 Shadow-Attack Aware Checkpoints

The defense described is applied to five bioassays by simulating the assay using the open-source UCR Static Simulator [65]. To emulate the high density of microelectrodes in a MEDA biochip, we broke up each DMFB electrode into 2×2 microelectrodes [76, 83]. Since these assays involve the mixing of two unit droplets, the largest droplet is of 4×2 size. We consider a MEDA biochip with a square microelectrode of 100 µm side. The average speeds of 4×2 and 2×2 sized droplets are 1 mm/s and 1.3 mm/s, respectively [81]. We use these parameters to analyze the susceptibility of the bioassay executions and to estimate the cost of the defense. Next, we drop the safe cycles that are not susceptible to any attacks from the CP list. This is referred to as the pruning mechanism (Table 3.2).

Table 3.2 exhibits the results of the analysis of the shadow attacks and the defense for five real-life bioassay implementations. Our evaluation reveals that all bioassays are vulnerable to shadow attacks. The recommended defense increases the memory storage expense by less than 7% (to store the golden droplet maps). The overhead is shown to be insignificant with CP pruning. The bioassays have numerous safe cycles that are not susceptible to any of the attacks, which can be dropped from the checkpoint list. This leads to a reduction of more than 45% in the number of checkpoints. This shows that the proposed defense is very practical and extensible.

Further, we applied the shadow attack analysis to three smaller sample preparation bioassays. The analysis reveals that these smaller bioassays are more prone to shadow attacks. These bioassays are targeted for smaller biochips compared to earlier considered bioassays. For example, the Remia sample preparation bioassay is targeted for 5×5 biochips and takes only 40 cycles to complete [73]; whereas, PCR bioassay is targeted for 13×15 biochip and completes in 969 cycles. This means that for these smaller bioassays, the droplets are in closer proximity, leading to more chances of shadow attacks. The defense increases memory storage by more than 70%. The pruning mechanism only has a marginal effect on the overall number of checkpoints.

Table 3.2 Shadow attacks and defense on real-life benchmarks

Baseline		Extra CPs for shadow defense			CP list pruning		
Assay	#CPs	Vulnerable cycles	#CPs	Storage overhead	#Safe cycles	#CPs	Storage overhead
PCR	969	29	998	3.0 %	659	339	−65%
InVitro 1	1317	33	1350	2.5 %	792	558	−57.6%
InVitro 2	3382	124	3506	3.7 %	1775	1792	−47%
Protein 1	5440	305	5745	5.6 %	3022	2723	−49.9%
Protein 2	24,151	1648	25,799	6.8 %	12,719	13,470	−44.2%
Remia	34	25	59	73.5 %	7	52	52.9%
Dilution	36	30	66	83.3 %	5	61	69.4%
PCR stream	70	70	140	100 %	0	140	100%

Chapter 4
Tools for Security

In the previous chapter, we described how an improved architecture can improve biochip system security. However, architecture improved, such as integrated sensors, increases fabrication cost, which makes it inapplicable for low-cost products. Here, CAD tools can be used to assist the designer in making optimal design choices. These tools enable the designer to weigh the design choices rather than make uninformed decisions. To this end, we present two tools in this chapter: 1. verification of checkpoint-based defense and 2. ML-based checkpoint monitoring.

4.1 Checkpoint-Based Security

Checkpoint Defense Recall that the DMFB can be equipped with sensors that allow us to monitor its state. The sensors (CCD camera) determine the presence of droplets and their volumes—eventually providing the means for defense against security attacks, as illustrated in Fig. 1.6. The defender (biochip designer) generates checkpoints from the golden actuation sequence. These are then compared against the results obtained by real-time sensors to validate the execution of the bioassay.

If all the droplets in the DMFB are monitored at all the time-steps, then the DMFB implementation will be secured [86]. This is due to the one-to-one mapping between the real-time DMFB snapshots and the bioassay. However, due to resource constraints affecting sensing and image processing, continuous real-time monitoring is not possible [121, 123]. Accordingly, defense mechanisms that rely on *checkpoints* have been proposed. The main idea is to explicitly monitor certain cells at certain time-steps (checkpoints), which satisfy the time constraints imposed by sensing and image processing. The aim is to choose the checkpoints such that they

---

Based on "How secure are checkpoint defenses in DMFBs?", TCAD, 2021 [115].

S. Mohammed et al., *Security of Biochip Cyberphysical Systems*,
https://doi.org/10.1007/978-3-030-93274-9_4

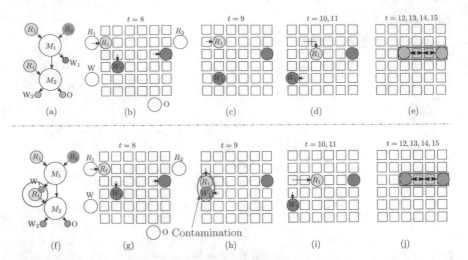

Fig. 4.1 Contamination attack demonstration on a DMFB. (**a**) Directed acyclic graph (DAG) of the bioassay. (**b**)–(**e**) Snapshots of the golden execution during time-steps $t = 8, 9, \cdots, 15$. (**f**) Modified DAG due to the actuation tampering. (**g**)–(**j**) Proximity attack in progress where the routing paths of droplets W_1, R_1 are modified between time-steps $t = 9$ and $t = 11$

provide enough coverage to make it difficult (ideally, impossible) for an attacker to tamper with the DMFB implementation.

Motivational Example The bioassay in Fig. 4.1a can be compromised by actuation tampering. The tampered actuation alters the route of droplets R_1 and W_1 between time-steps $t = 9$ and $t = 11$ to contaminate the droplet R_1 before the mix–split operation M_2. This yields a droplet with a different mixing ratio instead of R_1 : $R_2 = 3 : 1$, as shown in Fig. 4.1. If the DMFB snapshots are not monitored in time-steps $t = 9$ and $t = 10$ (Fig. 4.1e–f), the attack evades detection.

Defense Constraint Existing solutions have a spectrum range from "provably-secure-defense" that overshoots the available sensing resources to "probably-secure-defense" that works with the available sensing resources. Provably-secure-defense guarantees the valid bioassay execution on the DMFB [86]. However, this requires additional resources in terms of integrated sensors to monitor all the droplets at all the time-steps. On the other hand, probably-secure-defenses include randomized checkpointing, which checks random cells at random time-steps as well as static checkpointing, which checks the cells in the neighborhood of the droplet paths [121, 123]. These defenses only provide probabilistic guarantees against attacks. They do not consider all possible maneuvers an attacker can perform to avoid the checkpoints [92]. This opens the door for smart manipulations by an attacker to escape monitoring.

4.2 Exact Analysis of Checkpoint-Based Security

Given the large space of bioassay designs and the attacker's manipulation ability, it is not clear how secure the defenses are when applied to an arbitrary bioassay. Hence, there is a need to develop an exact methodology to analyze whether a given defense based on checkpoints indeed prevents the execution of an attack. First, we propose a methodology to determine if an actuation sequence can execute an attack without being detected by any checkpoint. We do so by considering *all* possible actuation sequences by means of a *symbolic formulation*. Using the symbolic formulation, we check whether at least one sequence exists which:

1. Can be implemented on the DMFB
2. Matches the original bioassay at all checkpoints
3. Attacks in the time-step not covered by the checkpoints

If such an attack is possible, the symbolic formulation yields an attack plan explicitly showing how the defense can be compromised. If no such sequence exists, the defense has been proven to be secure. Since resolving the proposed symbolic formulation is a complex task (eventually, this requires the consideration of all possible actuation sequences), we utilize the deductive power of satisfiability solvers. In this section, we outline the formulation and illustrate how this formulation can be solved using satisfiability solvers. Eventually, this allows assessing the strength of a checkpoint-based defense against attacks on a DMFB (Sect. 4.3.2). Moreover, this formulation can be used by the designer to devise a provably-secure checkpointing (Sect. 4.3.3).

4.2.1 Symbolic Formulation

Consider the bioassay DAG synthesized to the actuation sequence for the target $r \times c$ DMFB. The bioassay requires T time-steps to finish. We decompose the input bioassay into several edge-disjoint "input to output/waste" paths. Each path is a droplet trajectory that appears on the DMFB from an input reservoir (by dispense operation), takes part in one or more operations (e.g., mixing/detection), and is dispensed to a waste/output reservoir. For example, we can decompose the sequencing graph from Fig. 4.1a into three edge-disjoint paths: $(R_1 \rightarrow M_1 \rightarrow M_2 \rightarrow O)$, $(R_2 \rightarrow M_1 \rightarrow W_1)$, and $(R_1 \rightarrow M_2 \rightarrow W_2)$. Finally, we assign a unique identifier to each droplet and assume that the droplets appearing on the DMFB (i.e., their identifiers) are stored in a set $\mathcal{D}$. Then, the following variables are used to describe all possible sequences (for $1 \leq x \leq r$, $1 \leq y \leq c$, $d \in \mathcal{D}$, and $0 \leq t \leq T$):

$$
a_{x,y,d}^t = \begin{cases} 1, & \text{if a droplet } d \text{ appears on a DMFB cell } (x, y) \\ & \text{at time } t \\ 0, & \text{otherwise} \end{cases} \tag{4.1}
$$

Furthermore, variables are introduced for input/output operations (for $1 \le x \le r$, $1 \le y \le c$, and $0 \le t \le T$):

$$
ip_{x,y}^t = \begin{cases} 1, & \text{if a droplet is dispensed on a DMFB cell } (x, y) \\ & \text{at time } t \\ 0, & \text{otherwise} \end{cases} \tag{4.2}
$$

$$
op_{x,y}^t = \begin{cases} 1, & \text{if a droplet disappears from a DMFB cell } (x, y) \\ & \text{at time } t \\ 0, & \text{otherwise} \end{cases} \tag{4.3}
$$

4.2.2 Ensure Valid DMFB Execution

Solving this symbolic formulation (with free variables only) will admit arbitrary solutions and, hence, arbitrary actuation sequences. But sequences that violate obvious consistency constraints (e.g., a droplet suddenly appears in one-time-step and disappears in the next) have to be precluded. The following constraints are added to admit solutions that can be implemented on the DMFB:

(A) At time-step t, a droplet $d \in \mathcal{D}$ may appear in at most one DMFB cell:

$$
\bigwedge_{d \in \mathcal{D}} \bigwedge_{t=1}^{T} \left(\sum_{x,y} a_{x,y,d}^t \le 1 \right) \tag{4.4}
$$

(B) Each DMFB cell contains only one droplet in any time-step. The attacker can launch a malicious mix operation by transporting two droplets to a single cell.[1]

$$
\bigwedge_{t=1}^{T} \bigwedge_{x=1}^{r} \bigwedge_{y=1}^{c} \left(\sum_{d \in \mathcal{D}} a_{x,y,d}^t \le 2 \right) \tag{4.5}
$$

(C) The movements of droplets on the DMFB have to satisfy the following constraints. If a droplet $d \in \mathcal{D}$ is on a cell (x, y) at time-step t (i.e., $a_{x,y,d}^t = 1$), then:

---

[1] Since droplets greater than $2\times$ cannot be moved, we ignore them [56].

(a) Either d was on the same cell (x, y), or on one of its four neighbors (denoted by $N_4(x, y)$) in time-step $t - 1$.
(b) Or d is next to a dispenser creating the droplet on (x, y) at time-step t (this only needs to be described for locations (x, y), where droplets can be dispensed).

$$a^t_{x,y,d} \implies \underbrace{\left(\bigvee_{\substack{(x',y') \in \\ N_4(x,y) \cup \{(x,y)\}}} a^{t-1}_{x',y',d} \right)}_{(a)} \vee \underbrace{\bigvee ip^t_{x,y}}_{(b)} \qquad (4.6)$$

(D) Droplets may disappear when they leave the DMFB through a sink. Hence, if a cell (x, y) was occupied by $d \in \mathcal{D}$ at time-step $t - 1$, which is not present on its current location (x, y) or the neighborhood $N_4(x, y)$ at time-step t, then it must go to the sink on (x, y).

$$a^t_{x,y,d} \wedge \neg \left(\bigvee_{\substack{(x',y') \in \\ N_4(x,y) \cup \{(x,y)\}}} a^{t-1}_{x',y',d} \right) \implies op^t_{x,y} \qquad (4.7)$$

4.2.3 Ensure Expected Behavior at Checkpoints

In time-steps designated as checkpoints, the original behavior has to be maintained. This is ensured by adding the following constraints:

(A) We know the time-steps when droplets appear on the grid from an input reservoir. Let the k droplets $\mathcal{D}_{(x,y)} = \{d_1, d_2, \cdots, d_k\} \subseteq \mathcal{D}$ be dispensed on (x, y) at time-steps $t_{(x,y)} = \{t_1, t_2, \cdots, t_k\}$, where d_i is dispensed at t_i, for $i = 1, 2, \cdots, k$. The next constraint enforces correct dispensing for the input reservoir that dispenses droplets on location (x, y):

$$\bigwedge_{t \in t_{(x,y)}} ip^t_{x,y} \wedge \bigwedge_{t \notin t_{(x,y)}} \neg ip^t_{x,y} \qquad (4.8)$$

(B) Ensure the correct number of droplets disappear at each output reservoir. This prevents attacks due to the extra droplet [121].
(C) The designer places checkpoints to detect the presence or absence of a droplet. Let $\mathcal{CP}$ be the set of checkpoints. Each element in the $\mathcal{CP}$ is of the form (x, y, t, d_{size}), i.e., a droplet of size $d_{size} \in \{0, 1, 2\}$ must appear on (x, y) at t.

$$\bigwedge_{(x,y,t,d_{size})\in\mathcal{CP}} \left(\sum_{d\in\mathcal{D}} a^t_{x,y,d} = d_{size} \right) \tag{4.9}$$

4.2.4 Enforce an Attack

In this section, we model attack behaviors reviewed in previous chapters.

Deviation attack: Let each droplet $d \in \mathcal{D}$ have a defined location (x, y) for each time $t \in T$. This attack is modeled as follows:

$$\bigvee_{d\in\mathcal{D},t\in T} (\neg a^t_{x,y,d}) \tag{4.10}$$

This constraint captures any deviation from the synthesized bioassay. However, not all deviations from the golden actuation sequence lead to incorrect assay execution.

In the remaining part, we describe two classes of attacks, namely swap and proximity attacks (however, several other attacks can be modeled similarly). Without loss of generality, let us assume the DMFB supports dispense, transport, balanced mixing, and splitting. The proposed exact analysis ensures correct behavior at dispense operations by enforcing constraint (4.8), i.e., thwarts malicious dispense operations [121]. Moreover, if an attacker splits a unit-sized droplet into two halves, the child droplets cannot be transported further [33]. This malicious behavior, i.e., undesired splitting, can be detected by checkpoints placed on the droplet trajectory. Therefore, a split operation is not meaningful without a prior mix. The proximity attack, i.e., undesired droplets coming to close to each other, models the possibility of contamination and malicious (extra) mix operations. Further, the attacker can change the droplets in a mixing operation, as modeled by the swap. By ensuring that the given checkpoint thwarts these attacks, a correct assay execution is guaranteed. More precisely:

Swap attack: After dispensing from an input reservoir, a droplet can mix with one or more droplets before going to an output/waste reservoir. The swap attack swaps one of the input droplets of a mixing operation with an undesired droplet—or it transports a droplet to the wrong output location corrupting the bioassay output. In order to verify the feasibility of a swap attack, we need to check whether there exists an implementation of the DAG that satisfies the checkpoints but alters an input of the mixing operations or dispenses a droplet into the wrong output reservoir.

Suppose droplet $d \in \mathcal{D}$ is expected to be at the location (x, y) at time t. The swap attack is possible if d can be replaced by any other droplet $d' \in \mathcal{D} \setminus \{d\}$ without being detected by the checkpoint-based defense. The droplet d either can be an input droplet in the mixing operations ($\mathcal{M}$) or an output droplet. $\mathcal{M}$ is the set of all mixing operations, where each mixing operation is represented as $[(x_1, y_1, d_1, t_s), (x_2, y_2, d_2, t_s), t_{mix}, M_{type}]$, i.e., two $1\times$ droplets d_1 and d_2 come to the locations (x_1, y_1) and (x_2, y_2), respectively, at time-step t_s and after mixing using $M_{type} \in \{1 \times 4, 4 \times 1, 2 \times 4, \cdots\}$ during the next t_{mix} consecutive mixing cycles. The resulting two $1\times$ droplets (after balanced splitting) must come to the locations (x_1, y_1) and (x_2, y_2) at time-step $t_s + t_{mix}$. Similarly, the set of output operations $\mathcal{O} = \{(x_1, y_1, d_1, t_1), (x_2, y_2, d_2, t_2), \cdots, (x_k, y_k, d_k, t_k)\}$ denotes the droplet dispense locations and time. The swap attack is modeled by the next clause plus the checkpoint clause from earlier.

$$\bigvee_{(x,y,d,t)\in\mathcal{M}\cup\mathcal{O}} \left(\neg a_{x,y,d}^t\right) \tag{4.11}$$

Proximity attack: For the droplet located on cell (x, y) at time-step t, another droplet must come to any of the 8-neighboring cells of (x, y) (denoted $(N_8(x, y))$). This is formulated as follows:

$$proximity_{x,y}^t \Leftrightarrow \left(\left(\sum_{d\in\mathcal{D}} a_{x,y,d}^t = 1\right) \implies \bigwedge_{\substack{d\in\mathcal{D}, \\ (x',y')\in \\ N_8(x,y)}} \neg a_{x',y',d}^t\right)$$

$$proximity^t \Leftrightarrow \left(\bigvee_{\substack{1\leq x\leq r, \\ 1\leq y\leq c}} \neg proximity_{x,y}^t\right)$$

The Boolean variable $proximity_{x,y}^t$ is true if and only if no droplet is present in any of the eight neighbors of the location (x, y) at time instant t. Analogously, the variable $proximity^t$ is true if and only if there is a proximity attack at time-step t. Using these two Boolean variables, the following constraint enforces the proximity attack for a bioassay that requires T time-steps to finish.

$$\bigvee_{t=1,2,\cdots,T} proximity^t \tag{4.12}$$

4.2.5 Exact Analysis of a Checkpoint

The overall flow for the exact analysis of checkpoint-based defenses is summarized in Algorithm 1. The symbolic formulation represents arbitrary behavior on the considered DMFB. Adding the constraints that ensure a valid DMFB execution and the expected behavior at the checkpoints narrows the possible solutions that represent the "expected behavior." If adding the attack constraints leaves at least one satisfying solution, then the defense is not secure. To resolve the formulation, we use satisfiability solvers [53]. If the solver returns a satisfying assignment to all the variables used in the formulation for the particular bioassay, the defense is broken,

---

Algorithm 1 *IsAttackResilient*($\mathcal{A}, T, \mathcal{CP}$)

Input: $\mathcal{A}$: Actuation sequence of the bioassay; T: assay time; $\mathcal{CP}$: Set of checkpoints.

Output: Yes, if $\mathcal{CP}$ is attack resilient, otherwise, an attack as a counter-example.

Reverse engineer $\mathcal{A}$ to extract the DAG ($\mathcal{G}$) corresponding to the bioassay [35, 43] and other bookkeeping information such as dispense locations and mixer information.

Decompose $\mathcal{G}$ into edge-disjoint paths and assign a unique identifier to each path.

```
/* Detailed modeling - variables.                                    */
```
$M :=$ Add variables as defined in Eqs. (4.1)–(4.3).
```
/* Detailed modeling - constraints.                                  */
/* Add constraints for modeling valid DMFB operations.               */
```
$M :=$ Add constraints as given in Eqs. (4.4)–(4.7).
```
/* Add constraints for checkpoints.                                  */
```
$M :=$ Add constraints for enforcing input and output operations (Sect. 4.2.3) **for** *each checkpoint* $(x, y, t, d_{size}) \in \mathcal{CP}$ **do**

| $M :=$ Add constraints as given in Eq. (4.9).

end
```
/* Add constraints for attacks.                                      */
```
$M :=$ Add constraints for an attack such as proximity attack (Eq. (4.12)), swap attack (Eq. (4.11)), or deviation attack (Eq. (4.10)).

if *M is unsatisfiable* **then**

| **return** $\mathcal{CP}$ on $\mathcal{A}$ is attack resilient.

end

else

| **return** Satisfiable assignments as an attack on $\mathcal{A}$.

end

---

and an attack plan can be extracted from this assignment. The satisfying assignment to all $a^t_{x,y,d}$, $ip^t_{x,y}$, and $op^t_{x,y}$-variables gives the positions of all droplets, of all dispense operations, and of all disappearances of droplets, respectively, explicitly describing the plan for the attack. If the solver fails to return a satisfying solution, this is proof that there is no viable attack plan and, hence, the defense is secure. The following example shows how the exact analysis detects security vulnerabilities by returning an attack pathway as a counter-example.

Example 6 Let us consider the DAG for mixing two input reagents using *REMIA* [73], as shown in Fig. 4.2a. This bioassay implementation on a 5×5 DMFB takes $T = 40$ time-steps. Figure 4.2b–f show the snapshots of the DMFB for time-steps $t = 16, 17, \cdots, 20$. Checkpoints are incorporated by the designer as a defense and are highlighted on the DMFB grid in each time-step in Fig. 4.2. For these sets of checkpoints, the proposed exact analysis shows that two droplets can be swapped from time-step $t = 16$ to $t = 20$, as shown in Fig. 4.3b–f. The resultant modified bioassay is shown in Fig. 4.3a.

Suppose the designer incorporates a different set of checkpoints, as highlighted in Fig. 4.4a–e. Here, the cell $(5, 1)$ is not monitored at $t = 19$ compared to the checkpoint in Fig. 4.2. Now, the exact analysis shows that a proximity attack can be performed. It provides a pathway to the proximity attack that contaminates two droplets at time-step $t = 19$ as a counter-example (Fig. 4.4d).

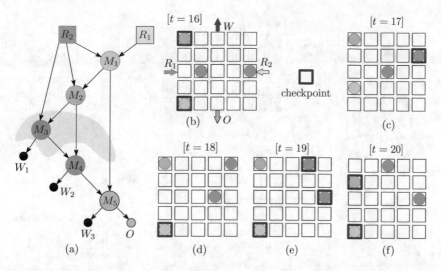

Fig. 4.2 (**a**) DAG for mixing reagents R_1 and R_2 using *REMIA* [73]. (**b–e**) DMFB snapshots that implement the highlighted portion of the DAG: (**b**) Snapshot after the mixing M_3 at time-step $t = 16$. (**c**)–(**f**) Droplet routing operations for subsequent assay operations. The checkpoints for monitoring are shown

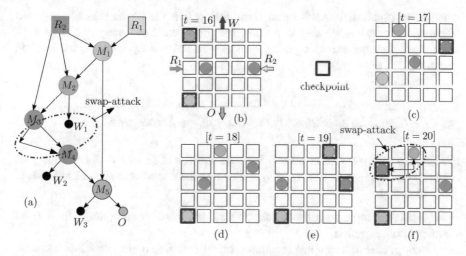

Fig. 4.3 (**a**) Modified DAG due to a swap attack. (**b**)–(**f**) Swap attack between the violet and green droplet ($t = 20$) is returned as a counter-example to prove that the checkpoints are incorrectly placed

4.3   Application of Exact Analysis Tool

The exact analysis has been implemented in Python 2.7 on an Intel Core-i7 machine. We use Z3 [53] to solve the resulting instance. In this section, we apply the exact

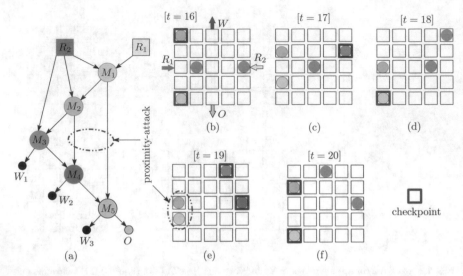

Fig. 4.4 (**a**) DAG showing the contamination attack. (**b**)–(**f**) Contamination attack (at $t = 19$) is returned as a counter-example to prove that the checkpoints are incorrectly placed

method to practical bioassays and show that it can verify the checkpoint-based defenses. Moreover, we also show how the proposed exact analysis can be used to derive a counter-example-guided provably-secure checkpointing determination method.

4.3.1 Considered Bioassays and Defense Strategies

We use the following set of DMFB implementations of bioassays (taken from [36] and representing practical use cases) to illustrate the importance of exact analysis in checkpointing:

- *REMIA*: a sample preparation scheme that minimizes reagent usage for a given target concentration.
- *Linear gradient*: a sample preparation scheme used to optimally dilute a sample in a linear gradient while minimizing wastage.
- *PCR mix*: a polymerase chain reaction (PCR), which is used for DNA amplification and involves mixing of seven fluids in the desired ratio.
- *PCR stream*: another PCR that is optimized to mix for multiple droplet generations as demanded by the application.

For each bioassay, a corresponding DMFB implementation is derived and summarized in Table 4.1.

Table 4.1 Considered bioassays from [36]

Assay	DMFB size	#Droplets	#I/O ports	Assay time (T)	#Mix–Split
REMIA	5×5	4	4	40	5
Linear gradient	9×8	6	5	40	6
PCR mix	8×15	8	10	70	7
PCR stream	8×15	13	11	73	15

Next, we applied a checkpoint (CP) defense, wherein all the droplets and input ports are checked at each time-step. This is referred to as *baseline defense*. The designer can derive multiple variants of baseline defenses by exercising the following options:

- Reduce the number of cells checked at each time-step.
- Increase the time interval between the checkpoints.

For the experiments considered here, several plausible decisions have been taken to this effect. These decisions are shown in the second and third columns of Tables 4.2 and 4.3.

4.3.2   Exact Analysis of Considered Defenses

In the following, we used the proposed exact analysis to evaluate the resulting CP strategies and analyze their security against swap, proximity, and deviation attacks. The current state of the art does not yet provide an exact analysis to verify whether those CP decisions indeed yield a secure chip. Through the method proposed in this chapter, we can now conduct such an analysis.

Tables 4.2 and 4.3 summarize the results obtained when exploring two variants of baseline defenses. More precisely, for each attack (Swap, Proximity, and Deviation) as well as for each considered number of monitored cells and checkpoint interval, it is listed whether the proposed analysis determines the defense as secure (denoted by P for pass) or insecure (denoted by F for fail). The first row for each bioassay (highlighted in yellow) lists the results obtained for the defense where all the droplets are checked at each time-step. However, the following rows list the results for the variants of the baseline defenses. Finally, the total run-time for conducting the checks is provided in the final column.

The results show that, for the first time, the proposed solution can be used as an oracle to evaluate the various options for the checkpoint-based defense design. This does not only allow to verify whether a particular defense is secure but also to trade-off, e.g., between "lowering the number of cells per checkpoint" versus "increasing the interval between checkpoints." The results suggest that the former is better compared to the latter. Further, the proposed method can be used by the designer for red-teaming against the defense, i.e., to learn from the counter-example

Table 4.2 Verify defense by varying the #cells checked at CP

Assay	CP interval	Cells /cycle	Secure? [Pass (P) / Fail (F)]			Run time
			Swap	Proximity	Deviation	
REMIA	1	4	P	P	P	0.21 s
		3	P	F	F	0.23 s
		2	F	F	F	0.3 s
Linear gradient	1	6	P	P	P	1.50 s
		5	P	P	F	1.3 s
		4	P	F	F	1.34 s
		3	F	F	F	1.02 s
PCR mix	1	8	P	P	P	7.78 s
		7	P	P	F	7.03 s
		6	P	F	F	7.46 s
		5	F	F	F	5.5 s
PCR stream*	1	11	P	P	P	15.95 s
		10	P	F	F	15.93 s
		9	P	F	F	12.87 s
		8	F	F	F	12.2 s

* In the PCR stream, the maximum number of droplets at a given cycle is 11.

Table 4.3 Verify defense by varying the interval between CPs

Assay	Cells / cycle	CP interval	Secure? [Pass (P) / Fail (F)]			Run time
			Swap	Proximity	Deviation	
REMIA	4	1	P	P	P	0.21 s
		2	P	F	F	0.15 s
		3	F	F	F	0.16 s
Linear gradient	6	1	P	P	P	1.50 s
		2	P	F	F	1.23 s
		3	F	F	F	1.37 s
PCR mix	8	1	P	P	P	7.78 s
		2	F	F	F	5.07 s
PCR stream*	11	1	P	P	P	15.95 s
		2	P	F	F	16.16 s
		3	P	F	F	16.16 s

* In the PCR stream, the maximum number of droplets at a given cycle is 11.

of a defense failure. The results show that exact security could be achieved by checking a lesser number of cells than the baseline solution, as shown by the *linear gradient* and *PCR mix* assays in Table 4.2.

One key observation from the results (Tables 4.2 and 4.3) is that *it is possible to guarantee the functional correctness of bioassay execution without monitoring all the electrodes at all time-steps*. In other words, it is possible to defend against swap and contamination attacks, thereby safeguarding the integrity of bioassay implementation. Note that without swap or contamination attack, the deviation attack does not violate the correctness of a bioassay.

4.3.3   Counter-Example-Guided Checkpoint Determination

Motivated by the results in Sect. 4.3.2, we additionally derived a provably-secure CP methodology that: (1) does not need to monitor all the electrodes at all time-steps and (2) is proved to be secure. Recall that at each checkpoint, the CCD camera is used to capture a snapshot of the DMFB at run-time. Then, the image is cropped into sub-images to extract the DMFB cell-of-interest. This sub-image is then correlated with a template of expected images that has predetermined information about droplet occupancy and size. Note that the number of time-steps covered by the CP strategy determines the number of times the CCD camera needs to capture the DMFB snapshots. Further, the number of image processing operations (sub-image extraction and correlation) is determined by the number of cells being in each of the CP time-steps.

In the proposed checkpointing, our objective is to determine a set of time-steps such that if all the droplet locations are monitored in these time-steps, then it ensures secure execution of the bioassay. We initialize the CP with the time-steps where mixing operations start. After that, we invoke exact analysis to check the security of the bioassay execution against different attacks for the given set of checkpoints CP. If the CP cannot secure the execution, the oracle returns a counter-example. We use this counter-example to analyze the point of defense failure and update the CP accordingly. The query and CP update continue until the CP guarantees the security of the bioassay execution. Figure 4.5 shows the overall scheme of a counter-example-guided secure checkpointing.

Fig. 4.5 Overview of secure checkpoint derivation

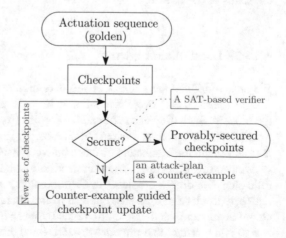

4.3.3.1 Checkpoint Time-Step Derivation

Recall that in a bioassay, a droplet's lifetime starts with a dispense operation, then it participates in assay operations such as mixing, incubation, and finally dispensed to the output/waste reservoir. We use the exact analysis oracle to verify if each droplet meets the expected behavior in each checkpoint (CP), i.e., each droplet participates in the desired assay operations and does not contaminate other droplets. The verification fails either due to a droplet d not participating in an operation op at time-step t or droplet d contaminating another droplet d' at time-step t. In such a case, the exact analysis oracle returns a counter-example. Then, we backtrace the droplet d's route from time t and place a new checkpoint at time-step t_1 where the droplet d begins to deviate from its specified route, leading to the verification failure at time-step t.

After updating the CP list, the oracle is run again to verify the security. Here, we use the incremental solving ability of the Z3 solver to reduce the run-time overhead between successive calls of the oracle, wherein the constraints for modified checkpoints are added as new constraints. This helps Z3 solver to decide the truth value of the instances quickly. This updating of CP and incremental solving is repeated until the coverage of the CP list is large enough to secure the bioassay execution. The pseudo-code for CP methodology guided by the counter-example is shown in Algorithm 2. In the following, respective details are described:

Example 7 Consider the bioassay shown in Fig. 4.6. The $\mathcal{CP}$ is initialized with the start time-steps ($t = 4, 12$) of two mix–split operations. With this set of $\mathcal{CP}$, the exact analysis returns a counter-example shown in Fig. 4.6. Here, droplet R_1 and W_1 are swapped, as shown in Fig. 4.6d. Algorithm 2 detects these are bad droplets, and backtracks their paths. The first deviation of droplet R_1 (W_1) droplets is at time-step $t = 9$ ($t = 10$). Time-steps $t = 9$ and $t = 10$ are added to the $\mathcal{CP}$, and the exact analysis continued. The updating of $\mathcal{CP}$ is continued until the exact analysis returns the status of $\mathcal{CP}$ as safe.

4.3.3.2 Local Minimization

We next minimize the number of droplets monitored in a CP time-step by doing a local search. Let a droplet monitored in a CP time-step be in a sparse area of the biochip. Such droplet can be safely dropped from the monitoring (CP) list; if the time interval between the current CP time-step (t) and the previous CP time-step (t') is smaller than the Manhattan distance between itself at t and its nearest neighbor at t'. We use this observation to minimize the number of droplets monitored in a CP time-step. We drop a droplet d in CP time-step $t \in \mathcal{CP}$ if the Manhattan distance between droplet d at time-step t and its nearest neighbor d' in the previous CP time-step t' is greater than the time difference between the time-steps ($t - t' + c$), where $c > 0$ is a constant that provides a guard-band. This way, the droplets that are far

---

Algorithm 2 *CP-Time-Steps*($\mathcal{A}$, T)

Input: $\mathcal{A}$: Actuation sequence of the bioassay; T: assay time;
Output: $\mathcal{CP}$: Set of checkpoint time-steps.
```
/* Initializing CP                                                          */
CP = start time t of all mixing operations in the assay
/* loop until bioassay is secure.                                           */
while true do
    /* Generate counterexample CE                                          */
    CE = IsAttackResilient(A, T, CP) if CE == φ then
     |  return CP;
    end
    else
        /* Find droplets that cause verification failure                   */
        bad_droplet = [] for each droplet d do
            for each assay operation op do
                if d does not participate in op or d is contaminated then
                 |  Add (d, op, t) to bad_droplet
                end
            end
        end
        /* Update CP list                                                  */
        for (d, op, t) in bad_droplet do
         |  Backtrace first deviation point t₁ for d leading to op at t. Add t₁ to CP
        end
    end
end
return CP
```

---

from the other droplets are safely dropped from the CP list. This minimizes the number of cropping of sub-images and correlation performed in a CP time-step.

Example 8 Consider the case of Example 7, wherein time-steps $t = 9, 10$ are added to $\mathcal{CP}$ by the counter-example-guided checkpoint update routine. Here, M_1 droplet at $t = 10$ is farther from the other two droplets at the previous CP time-step ($t' = 9$) by at least three steps (Manhattan distance), as shown in Fig. 4.7b. The time difference between the time-steps is $t - t' = 1$. We choose $c = 1$, to avoid the possibility of droplets coming into the neighboring cell. This means that the droplet M_1 at $t = 10$ satisfies our sparsity condition. Therefore, the droplet M_1 at $t = 10$ is dropped from $\mathcal{CP}$. This avoids the computation of the image correlation of one electrode. However, the droplet R_1 at $t = 10$ cannot be dropped as the Manhattan distance between droplets at $t = 10$ from droplets at $t' = 9$ is smaller than the time difference of time-steps. Therefore, all the droplets at $t'' = 12$ are retained.

4.3.3.3 Experimental Results

We applied the proposed CP generation scheme to real-life benchmark assays described in Sect. 4.3.1. The results of the CP generation are tabulated in Table 4.4.

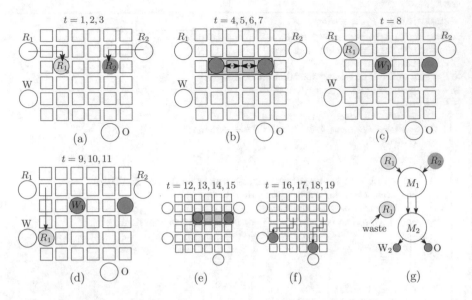

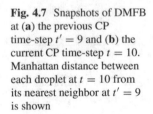

Fig. 4.6 Actuation tampering. (**a**)–(**f**) Swap attack in progress where droplets W_1, R_1 are swapped between $t = 8$ and $t = 10$. (**g**) Modified DAG due to the attack

Fig. 4.7 Snapshots of DMFB at (**a**) the previous CP time-step $t' = 9$ and (**b**) the current CP time-step $t = 10$. Manhattan distance between each droplet at $t = 10$ from its nearest neighbor at $t' = 9$ is shown

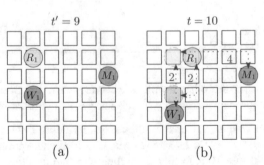

Table 4.4 Secure CP generation results

	Baseline		CP time-step derivation					Local pruning	
	#CP time -steps (τ_b)	#CP cells (c_b)	#CP time -steps (τ_s)	#CP cells (c_s) Assay	#iter	Run-time	$\frac{\tau_s}{\tau_b}$	#CP cells (c_l)	$\frac{c_l}{c_b}$
REMIA	40	112	16	55	15	66.6 s	40%	48	42%
Lin. gradient	40	204	16	105	11	73.7 s	40%	69	34%
PCR mix	70	368	35	229	30	1226 s	50%	182	49%
PCR stream	74	604	50	454	40	3557 s	66%	384	63%

The proposed CP time-step derivation scheme monitors between 40% (REMIA) and 66% (PCR stream) cells to provide provable security compared to 100% cell coverage requirement in baseline strategy. The total run-time varies from 1 min (REMIA) to 60 min (PCR stream), which is a one-time cost incurred offline during design. The number of cells to be monitored is further reduced from 34% (Lin. gradient) to 63% (PCR stream) by using local minimization. We used $c = 1$ for our experiments and verified the security through the verify routine. This shows that the utility of the proposed verification oracle in developing provably-secure-defense with practical resource requirements.

The total run-time for CP time-step derivation is determined by the number of invocations of the checkpoint verify routine. The run-time of the verify routine depends on the size of the bioassay and the DMFB size. We can reduce the run-time overhead of the exact verification by decomposing the input bioassay into multiple sub-bioassays and verify each sub-bioassay separately. This will ensure the security of the entire bioassay and yet make the verification process scalable.

4.4 Machine Learning for Attack Detection

Next, we propose an ML-based attack detection framework for checkpoint-based monitoring of biochips. We first discuss the motivation for using an ML model for attack diagnosis, and then we describe the data formats, ML models, and performance metrics used in the analysis.

4.4.1 The Motivation for Using ML in Attack Diagnosis

To better understand the attacker's capabilities and intent, the biochip defender needs to identify an attack. An attack classification helps the defender devise proactive security measures. The attack detection process is complex because: (1) the FPVA biochip offers a general-purpose microfluidic platform; therefore, the attacks are possible in a variety of forms and ways; (2) the input data has incomplete information of biochip state for the sensor resource limitation; (3) the analysis becomes time-consuming with the increase in biochip size and/or decrease in sensor coverage.

We can adopt randomized checkpointing [124] to detect an attack on a fully-programmable-valve-array (FPVA). However, the probability of evading detection of an attack P_E (Equation (3.1)) in the case of incomplete data corresponding to valve/node states can be as high as 100%. To improve the probability of detection $(1 - P_E)$ we investigate the use of a machine learning (ML) framework for attack classification. In other words, we wish to design a model that can learn the attack patterns from incomplete data. Our results show that the ML-based classifier not only detects attacks with better accuracy but also provides a diagnosis of the type of attack. The design flow of an ML-based classifier is shown in Fig. 4.8.

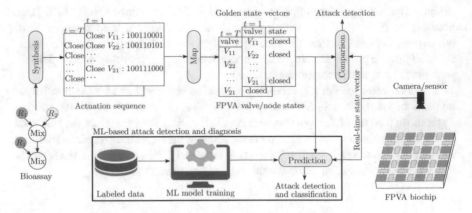

Fig. 4.8 The traditional attack detection flow (top row) and the proposed additional ML-based (inset) attack detection-cum-classification flow

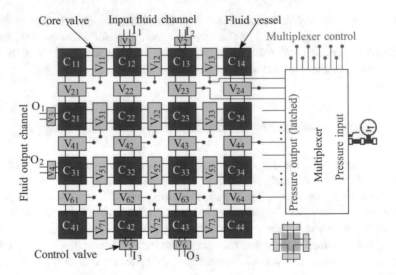

Fig. 4.9 Schematic of a fully-programmable-valve-array biochip

4.4.2   Data Format for FPVA

An FPVA biochip is a two-dimensional array of fluid chambers. Each chamber in an FPVA is surrounded by up to four independently addressable valves to implement programmable interconnection of chambers (Fig. 4.9). Designers can configure valves to create an arbitrary channel network connecting the desired chambers. The chambers are used as vessels to hold and mix reagents [127]. To scale down the number of I/O pins, FPVAs use a multiplexer to control each of the valves. This offers a versatile programmable platform to implement a broad range

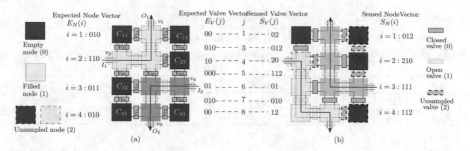

Fig. 4.10 Encoding of the state of the FPVA valves. (**a**) Expected node vector $E_N(i)$ and expected valve state vector $E_V(j)$ of ith and jth row. (**b**) Real-time sensed node state vector $S_N(i)$ and sensed valve state vector $S_V(j)$ of ith and jth row. The dashed nodes and dashed valves are not sensed

of bioassays [59]. The state of an FPVA biochip can be represented by a set of row vectors corresponding to the node and/or valve states in each row. The example below explains the format for a row.

Example 9 Consider an FPVA shown in Fig. 4.10a. The first row ($i = 1$) of nodes has one filled (C_{12})—shown in a light color and two empty nodes (C_{11} and C_{13})—solid blue color. This is represented by node state vector "010." Similarly, the first rows ($j = 1$) of two valves V_{11} and V_{12} are closed. This is represented by the valve state vector "00." The representation of the state of the entire FPVA biochip is discussed next.

The data set used for building ML models is composed of expected golden biochip state E, real-time sensed biochip state S, and the attack class label C. The expected golden state of the biochip is derived from the actuation sequence, as shown in Fig. 4.8. The real-time biochip state can be sensed through the sensor feedback in the biochip cyberphysical system. However, sensors (CCD camera) can only observe a limited number of valves of the biochip, due to the limited field of view. This is captured by the unknown or unsampled valve states in the analysis. We can capture the state of the biochip by sampling the nodes and/or valves in the following ways:

- An $m \times n$ FPVA biochip has $m \cdot n$ nodes. We match the cropped images of such nodes with a golden template to determine the state of the node, i.e., filled or empty. Note we represent the state of each node in the data set using 1 (filled), or 0 (empty), and 2 (unsampled). We group the node state representation corresponding to a row. Figure 4.10 shows the node state vectors for a 4×3 FPVA biochip. Table 4.5 shows a sample data set of node state vectors.
- An $m \times n$ FPVA has a maximum of $(2 \cdot m \cdot n + m + n)$ valves, which includes internal and I/O valves. We match the cropped images of such valves with a golden template to determine if the valve is open or closed. We represent the state of each valve in the data using 1 (open), or 0 (closed), and 2 (unsampled).

Table 4.5 Node state data set format for attack classification

The expected node state vector $E_N(i)$				The real-time node state vector $S_N(i)$				
$i=1$	$i=2$	$i=3$	$i=4$	$i=1$	$i=2$	$i=3$	$i=4$	Class
010	110	011	010	010	110	011	010	No-attack
010	110	011	010	010	110	111	110	Transpose
010	110	011	010	010	110	011	010	Tunneling

The valve state representation for each row is grouped together in a vector. Figure 4.10 shows the valve state data of the 4×3 FPVA biochip.

Example 10 Consider the FPVA platform of size 4×3 shown in Fig. 4.10. Figure 4.10a shows the golden implementation from which we derive the expected node state vector $E_N(i)$ for rows $i = 1, 2, 3, 4$ and expected valve states vector $E_V(j)$ for rows $j = 1, 2, \cdots, 8$. In Fig. 4.10b, a transpose attack swaps the two fluid paths. The real-time sensed node state vectors $S_N(i)$ for $i = 1, 2, 3, 4$ and valve state vectors $S_N(j)$ for $j = 1, 2, \cdots, 8$ corresponding to the transpose attack are shown in Fig. 4.10b. Table 4.5 shows node state vectors for various attack/no-attack scenarios following the convention explained earlier.

4.4.3 Data Acquisition

To generate data for the training of machine learning models, we consider several inputs to output paths on an $m \times n$ FPVA and exercise attacks on the configuration. We assume $2 \cdot (m+n)$ input/output ports on the boundary of the FPVA biochip. First, we choose several inputs and output pairs. Then, for each input and output pair, we considered various channel patterns. Through this process, we generated more than $100K$ channel configurations. Next, we simulated tunneling and transpose attacks on each of the configurations. This way, we generated unbiased $100K+$ data for three attack classes: no-attack, tunnel, and transpose. Table 4.5 shows a node data set composed of sampling on a 4×3 FPVA biochip. The state vector corresponding to the biochip in Fig. 4.10 is shown in the second row (transpose class) of Table 4.5.

4.4.4 ML Models

We used the data set generated in the previous section (Sect. 4.4-C) to train several ML models and study their accuracy in attack classification. We studied the following linear and non-linear classifiers [69, 98]:

1. Logistic Regression (LR): models the relationship between variables through iterations to minimize a measure of the error in the prediction. We used L2 regularization as we expect most of the coefficients to be non-zero. The classification is decomposed into the binary classifiers for each class in a one-vs-rest fashion.
2. Linear discriminant analysis (LDA): exploits inherent structure in the data to simplify n-dimensional data. It does not perform well when data is non-linear. The purpose of using LDA was to showcase the usage of this basic model for the attack diagnosis.
3. K-Nearest Neighbors (KNN): stores all available cases and classifies new cases based on a similarity measure (e.g., distance functions). In this chapter, we used the squared distance from the mean of the labeled data as a measure of similarity.
4. Classification and Regression Trees (CART): constructs decision trees based on attributes of the training data. We used a strategy to choose the "best" split along with maximum features.
5. Gaussian Naive Bayes (NB): applies Bayes' theorem for classification with the "naive" assumption of conditional independence between every pair of features given the value of the class variable.

We split the data randomly into a training data set (80%), and we held back (20%) of the data to test the accuracy of the classifiers. Algorithm 3 shows the training and validation steps in the ML framework. In this chapter, we use five basic ML models to showcase that ML-based attack diagnosis is more effective compared to randomized checkpointing (Fig. 4.8). Other ML models can be used to push the limit of the diagnosis accuracy.

Algorithm 3 ML-assisted classifier assessment framework

Input: data_file; model
```
/* Data set file and ML model                                      */
```
Output: Accuracy, F1 score, learning curve
$X, Y = $ read_csv(data_file)
```
    /* 'X' denotes features; 'Y' is the label                      */
```
$X_{train}, Y_{train}, X_{test}, Y_{test} = $ train_test_split(X, Y, test_size $= 20\%$)
```
    /* Data is split into train (80%) & test (20%)                 */
```
model.fit(X_{train}, Y_{train})
$Y_{predicted} = model.predict(X_{test})$
print(classification_report($Y_{test}, Y_{predicted}$))

4.4.5 Performance Metrics

We use the following performance metrics to quantify the performance of ML models [75].

Accuracy ($\mathcal{A}$) of a classifier measures how often it predicts correctly over all the kind of predictions. Formally,

$$\mathcal{A} = \frac{TP + TN}{TP + TN + FP + FN} \tag{4.13}$$

where *TP*, *TN*, *FP*, *FN* are true positive, true negative, false positive, and false negative predictions, respectively.

F1 score ($\mathcal{F}1$) is the harmonic mean of two important classifier correctness measures: precision and sensitivity, which are defined as follows:

$$\text{precision} = \frac{TP}{TP + FP} \qquad \text{sensitivity} = \frac{TP}{TP + FN}$$

$$\mathcal{F}1 = \frac{2 \cdot \text{precision} \cdot \text{sensitivity}}{\text{precision} + \text{sensitivity}} \tag{4.14}$$

In other words, $\mathcal{F}1$ is a weighted average of sensitivity and precision.

Learning Curves In the classification problem, the learning curve shows the improvement in performance with respect to the size of training data. This helps to determine the requisite data size for training the ML model.

4.5 Checkpointing Scheme

In this section, we discuss various trade-offs in checkpoint design and then describe a checkpointing scheme to boost the classifier's accuracy.

4.5.1 Trade-Offs

Programmable biochips are susceptible to tampering of the high-level control program and the low-level actuation sequence. These are called "actuation tampering" attacks [32, 121]. The actuation sequence encodes a sequence of fluidic operations such as dispensing, transporting, and mixing. Tampering of actuation sequence impacts the fluidic operations on the biochip. For an FPVA, we define three types of tampering (shown in Fig. 4.11): (1) **Transpose attack** modifies golden operations, e.g., the mixing ratio or swapping of flows. (2) **Tunneling attack** inserts extra actuations to create new flow channels to dispense malicious fluids or contaminate flows. (3) **Aging attack** subjects the valves that are not participating in certain bioassay stage(s) to additional actuations. This reduces the biochip lifetime by failing these valves.

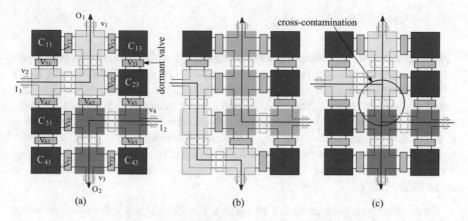

Fig. 4.11 FPVA attacks. (**a**) Original operation of routing two fluids. (**b**) A transpose attack. (**c**) A tunneling attack. Grey valves are susceptible to aging

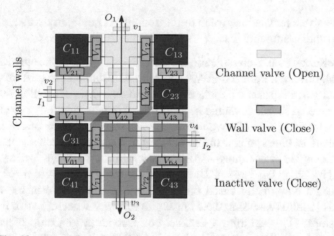

Fig. 4.12 Classification of the valves in a given configuration: channel valves, wall valves, and inactive valves

Cyberphysical FPVA platform uses a CCD camera to monitor bioassay execution, and it can monitor a subset of nodes and valves due to resource and computational constraints [121, 125]. The number of nodes is fewer than that of valves. Therefore, node monitoring gives better coverage of resources. However, node monitoring cannot detect aging attacks and tunneling attacks (when a single valve isolates two channels). Moreover, if the fluid flowing is colorless, it is difficult to differentiate a node state. Further, valve monitoring can detect aging and tunneling attacks.

4.5.2 Smart Checkpointing

Owing to the resource constraints, we can sense only a subset of nodes in each time-step. This leads to incomplete information being used by the machine learning (ML) algorithms while detecting and classifying an attack. Let us denote a "checkpoint" using (x, y, t). At time t, the checkpoint (x, y, t) monitors the valve/node state on (x, y) [92]. To boost the accuracy of the classifier algorithms, we propose a checkpointing scheme. The main idea is to sense around the channel path to increase the prediction accuracy of the ML classifier. We classify the valves into three types for an FPVA configuration. Figure 4.12 shows different valves in the configuration of an FPVA [127]:

- Channel valves: connect nodes to form a channel. These valves remain open during flow pushing.
- Wall valves: surround the channel to isolate different flows. These valves remain closed.
- Inactive valves: remain inactive in the current FPVA configuration/state but active in some other configurations.

Based on the above valve classification, we propose a checkpoint selection strategy shown in Algorithm 4. For each FPVA configuration/state, the algorithm differentiates between valve types and populates them into three lists depending on the valve type. For a given resource coverage k, Algorithm 4 selects randomly $c\%$ of coverage from channel valves, $w\%$ of coverage from wall valves, and $i\%$ of coverage from inactive valves as checkpoints. In the data set, the valves are represented by their golden and sampled state values—1/0. The rest of the valve states are unsampled—represented by "2" in the data set. The ratio of valve types being observed ($c : w : i$) can be selected empirically based on their level of importance in the context of attacks. This requires assessing their impact on ML model performance through the following steps: (1) generating a data set corresponding to a ratio of valve types being observed ($c : w : i$) and (2) retraining ML model with the data set. These two steps can be repeated over different $c : w : i$ until satisfactory performance is achieved.

4.6 Experimental Results

In this section, we evaluate the performance of ML-assisted attack diagnosis. We used the *Scikit-LearnML* tool (v0.22) [104] in *Python 2.7* to build our ML model for data analysis. We considered an 8×8 FPVA platform and performed two sets of experiments to (1) find the best fit ML-based classifier and (2) study the trade-offs in checkpointing design.

Algorithm 4 Smart checkpointing

Input: Golden actuation sequence, coverage (k), assay time T, valve type ratio $c : w : i$ (channel
          valve : wall valve: inactive valve)
Output: Set of checkpoints
array actuate[m][n][t];

List $chanel\_lst, wall\_lst, inact\_lst$;

for $(t = 1; t \leq T; t = t + 1)$ **do**
    for $(x = 1; x \leq M; x = x + 1)$ **do**
        for $(y = 1; y \leq N; y = y + 1)$ **do**
            if $actuate[x][y][t]=1$ **then**
                add (x, y, t) to $chanel\_lst$

            end
            else if $neighbor\ of\ actuate[x][y][t]=1$ **then**
                add (x, y, t) to $wall\_lst$

            end
            else if $(x, y, \tau)\ in\ chanel\_lst\ for\ \tau! = t$ **then**
                add (x, y, t) to $inact\_lst$

            end
        end
    end
    List $a =$ choose $(c * k)$ from $chanel\_lst$
    List $b =$ choose $(w * k)$ from $wall\_lst$
    List $c =$ choose $(i * k)$ from $inact\_lst$
    Set checkpoint on each location of $a \cup b \cup c$ at t

end

4.6.1 Assessing ML Classifiers

Using the data set described in Sect. 4.4-D, we studied the five different classification algorithms mentioned in Sect. 4.4-E. We calculate the accuracy $\mathcal{A}$ and
$\mathcal{F}1$ score for the data set pertaining to nodes and valves, as shown in Table 4.6.
Note we assumed full sensor coverage in this set of experiments. Therefore, the
values in each row of Table 4.6 indicate the best performance achievable by the
classifier. The experimental results confirm that the sensing of valves gives better
performance compared to node sensing. The learning curve shown in Fig. 4.13
ensures the adequacy of the data set size (300,000+) for training. "Classification and
regression trees (CART)" gives marginally better performance compared to the rest,
with the accuracy of prediction 0.97. Even the basic model of "Linear discriminant
analysis (LDA)" gives an accuracy of 0.71 for valve-data-set, which is better than
the average probability of attack detection $(1 - P_E)$ achieved using traditional non-
ML methods [94, 124]. The design cost in terms of time taken to fit each model and
the memory size of each model is summarized in Table 4.7. The accuracy of the ML

Table 4.6 Performance of classifiers with 100% coverage

Algorithm	Node data set		Valve-data-set	
	Accuracy $\mathcal{A}$	$\mathcal{F}1$ score	Accuracy $\mathcal{A}$	$\mathcal{F}1$ score
LR	0.62	0.57	0.87	0.87
LDA	0.62	0.55	0.71	0.67
KNN	0.53	0.53	0.93	0.93
CART	0.53	0.53	0.97	0.97
NB	0.76	0.75	0.44	0.31

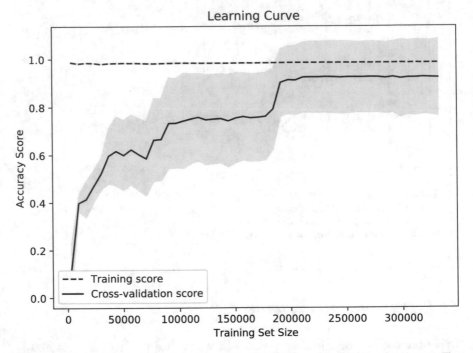

Fig. 4.13 The learning curve for the training of the CART ML model using the valve data set. The solid line indicates the mean accuracy score, and the shaded region indicates the standard deviation of the score

Table 4.7 Classifier design cost for valve-data-set

Algorithm	LR	LDA	KNN	CART	NB
Run-time (sec)	34	2.9	8.84	4.09	2.05
Memory (KB)	2.2	4.4	150,817	3624	2.9

models can be further improved by using more advanced models such as support vector machines, boosting methods, etc. This is left for future work.

4.6.2 Smart Checkpointing Assessment

We assess the impact of smart checkpointing compared to random checkpointing. Without loss of generality, we choose the CART classifier for this study. However, the subsequent analysis holds true for other ML models. We vary the sensor coverage (k) from 1/8-th fraction of all valves (114 valves) to 8/8 (full coverage of valves). Further, to determine the best ratio of valve types ($c : w : i$), we compare the performance of the ML model with different ratios of valve types used for smart checkpointing. Table 4.8 summarizes the results.

Analysis of ML model performance shows that the CART ML model fits well in the case of incomplete valve state information. With the coverage of 50% ($k = 4/8$), 0.95 accuracy can be achieved, which is just 0.02 less than the maximum accuracy possible (with 100% coverage). Further, the best performance is observed when the ratio of valve type coverage is $c : w : i = 50 : 50 : 0$, i.e., of all the observed valves, 50% are channel valves, and 50% are wall valves. This shows that the channel valves and the wall valves are easy targets for an attacker. Any change in these valves would cause a malfunction in the bioassay. In contrast, inactive valves can be used to degrade the biochip lifetime (aging attack). If the defender does not observe any inactive, then it is not possible to detect the aging attack. Therefore, a practical strategy could be to observe fewer inactive valves and more channel and wall valves. This could ensure a balance between the accuracy of prediction among all types of attacks.

Table 4.8 ML model performance for varying coverage k for random and smart checkpointing schemes. Smart checkpointing scheme performance with various ratios of valve type being observed $c : w : i$ (channel valve: wall valve: inactive valve)

| | Accuracy ($\mathcal{A}$) | | | | $\mathcal{F}1$ score | | | |
| | Random | Smart CP | | | Random | Smart CP | | |
Coverage	CP	40 : 40 : 20	45 : 45 : 10	50 : 50 : 0	CP	40 : 40 : 20	45 : 45 : 10	50 : 50 : 0
1/8	0.32	0.32	0.33	0.39	0.32	0.33	0.34	0.37
2/8	0.40	0.44	0.50	0.56	0.39	0.45	0.50	0.55
3/8	0.45	0.57	0.63	0.73	0.44	0.58	0.64	0.73
4/8	0.53	0.71	0.80	0.87	0.52	0.71	0.80	0.87
5/8	0.62	0.83	0.89	0.95	0.62	0.84	0.90	0.95
6/8	0.78	0.92	0.94	0.96	0.78	0.92	0.94	0.96
7/8	0.90	0.95	0.96	0.97	0.90	0.95	0.96	0.97
8/8	0.97	0.97	0.97	0.97	0.97	0.97	0.97	0.97

Chapter 5
Watermarking for IP Protection

A bio-protocol can be targeted for a biochip-based point-of-care diagnostic platform [6, 101]. An attacker can gain a biochip product from the marketplace and reverse engineer the bioassay [43]. Alternatively, a bio-protocol can be targeted for personalized drug development [90]. In such experiments, the biochip controller is connected to the network for round-the-clock online monitoring and control [17]. The attacker can access the sensor data (e.g., biochip images) through a network attack [96] and perform image analysis to reverse engineer the bioassay from the actuation sequence [43]. The need for protecting the IP can be well understood through the laborious bio-IP development process.

5.1 Bio-protocol IP Development

Here, we present a study of bio-protocol IP development from their benchtop description to their DMFB adaptation.

5.1.1 Benchtop Bio-protocol

In general usage, the term *protocol* refers to an exact formulation of the sequence of steps in a procedure, but without adding any interpretation. In biology, a protocol (referred to as bio-protocol) is used as a guideline that specifies ingredients (material and reagents), equipment, and the sequence of steps for performing an experiment. In molecular biology, the written protocol descriptions have a proprietary purpose

---

Based on "Bioprotocol Watermarking on Digital Microfluidic Biochip", TIFS 2019 [114].

© The Author(s), under exclusive license to Springer Nature Switzerland AG 2022
S. Mohammed et al., *Security of Biochip Cyberphysical Systems*,
https://doi.org/10.1007/978-3-030-93274-9_5

of claiming ownership. Similar to a culinary recipe, a bio-protocol often sketches an outline of essential steps, leaving considerable discretion about its actual realization on any particular platform. Benchtop implementation of a protocol is based on prior experience with a similar procedure and the designer's choice of ingredients, selection of parameter values, and specialized practices.

Example 11 An immunoassay is a ubiquitous technique to quantify molecules of biological interest (proteins, nucleic acids, and pathogens) based on the specificity of antibody–antigen interactions [80]. The success or failure of the immunoassay protocol depends on several factors such as immunoassay type (sandwich, competitive, or non-competitive), reagents, detection method (colorimetric, fluorescence, or chemiluminescence), sensitivity (pg/ml, ng/ml or μg/ml), and curve-fitting model for calibration curves used for quantification [51]. Only after several experiments can a protocol developer understand the interaction between these variables and determine appropriate parameter values that favor the desired outcome, e.g., specificity over yield, or sensitivity over specificity.

5.1.2 Adaptation to Microfluidic Platform

The advancement of microfluidic technologies enables an automated and low-cost platform for implementing benchtop bio-protocols [11, 13]. The benchtop to microfluidic platform adaptation has three highly interdependent components: setting outcome objectives, microfluidic pathway design, and determining valid parameter range, as described below:

5.1.2.1 Outcome Objectives

The developer sets the outcome objectives of the bio-protocol, such as execution time, output quality, and reagent wastage. These objectives influence the DMFB pathway choice and parameters.

Example 12 "Ella" immunoassay bio-protocol platform has multiple objectives, which include $\leq 10\%$ coefficient of variation (CV) in bio-protocol results, $\leq 25\,\mu$L sample volume, high reproducibility, and ≤ 1 h processing time [12].

5.1.2.2 Pathway Design

In a DMFB implementation of bio-protocols, the developer re-imagines the operations in the benchtop bio-protocol to find a DMFB-specific pathway.

Example 13 The microfluidic implementation [99] of the immunoassay protocol differs substantially with respect to its benchtop implementation [16]. In the benchtop implementation, centrifuge operations are used to wash, separate, and concentrate biological materials. This operation is implemented on the DMFB by using bio-activated magnetic beads, and the centrifugal operation is mimicked by using magnets.

5.1.2.3   Parameter Range

After determining the DMFB-compatible pathway, the developer needs to choose appropriate values for parameters to attain the desired objectives. This requires a systematic understanding of the interplay of such parameters [19], which include mixing time, incubation time, mixing ratio, reagent volume, and concentration. The parameters have a range of acceptable values due to the inherent variability in the bio-protocol [37]. The developer, through many trials, determines the parameter variable range. Each parameter is dependent on the design pathway, the outcome objective, and other parameter values. For example, three different particle-based immunoassay bio-protocols [99, 118, 132] use different bead sizes and different carrier fluids, resulting in different performance outcomes. This showcases the complex relationship between the three bio-protocol IP components: design pathway, parameter values, and the desired outcomes. The complexity of analysis is further increased due to the sheer number of choices that the developer has to make. Therefore, the developer invests high effort and cost in analyzing the interplay of these bio-protocol IP components.

Example 14 A developer of immunoassay bio-protocol is faced with multiple design choices, such as a method of antibody immobilization (physical adsorption or covalent attachment), a method of antigen delivery and washing (grooves in microchannels or active electric and magnetic forces), a type of platform (microarrays, magnetic or non-magnetic bead), a type of fluid handling (continuous flow or digital microfluidic), and detection strategy (surface plasmon resonance, fluorescence, or electrochemiluminescence) [102, 136]. Each of the design choices has pros and cons that are studied experimentally.

5.1.3   Biochip Cyberphysical System

The DMFB is susceptible to various imperfections and faults, which adds to the inherent variability in the bio-protocol implementation. To overcome this variability and for error recovery in DMFB operations [84, 140], real-time monitoring through sensor feedback is required, as shown in Fig. 5.1. In a bio-protocol implementation, the feedback control path structure is hierarchical: global and local. At the global level, the bio-protocol is divided into sub-protocols. The output of each sub-protocol is verified against quality criteria; based on this verification, the relevant sub-protocols are repeated if the output quality is not in the calibrated range, as shown

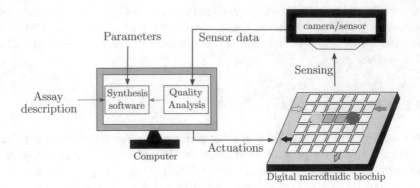

Fig. 5.1 A cyberphysical biochip system with feedback control

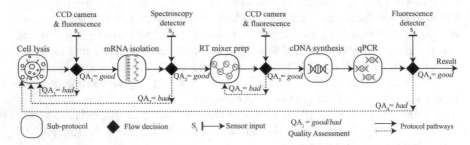

Fig. 5.2 A quantitative protocol control flow for gene expression analysis on a cyberphysical DMFB [74]. The protocol consists of five sub-protocols: cell lysis, mRNA isolation, RT mixer preparation, cDNA synthesis, and qPCR. The cyberphysical DMFB uses sensor feedback to assess the quality of each sub-protocol output and direct the control flow of the protocol

in Fig. 5.2. Similarly, at the local level, each sub-protocol consists of a sequence of fluidic operations, and the output of intermediate operations is monitored to verify the volume and concentration of the droplets.

Example 15 Quantitative gene expression analysis is used for disease diagnostics and pathogen detection. In this bio-protocol, uncontaminated cells are lysed, and nucleic acids are isolated for cDNA synthesis. After reverse transcription of cDNA, real-time quantitative PCR (qPCR) is performed for the amplification of the target gene for quantitative analysis. The mapping of this protocol onto the DMFB platform requires cyberphysical integration for dealing with the inherent uncertainty of the quality of cell lysis, mRNA extraction, and reverse transcription mixture [74]. Figure 5.2 shows the control flow of the DMFB implementation of quantitative gene expression analysis. Sensor readings are calibrated depending on the sensor type (optical, CCD-based) and target application [132].

5.2 Watermarking-Based IP Protection

We consider the scenario where a bio-protocol laboratory invested large sums of money and person-hours in developing a bio-product and wants to protect their intellectual property. It motivates a competitor to steal the IP and market it by incurring no cost for research and development. The bio-protocol developer tries to avert this by inserting a watermark into the bio-protocol. By using watermarking, we can embed secret information in the design as proof of ownership. When forgery is suspected, the secret information is retrieved from the misappropriated IP to serve as undeniable authorship proof [106].

5.2.1 Bio-protocol IP Protection

Bio-protocol IP development is a multi-phase process, as shown in Fig. 5.3. A robust IP protection mechanism is required to protect the innovations in each step and the bio-protocol development as a whole. We propose a hierarchical watermarking technique described as follows:

Registration: The DMFB-specific pathway derivation requires a reinvention of benchtop protocols. This can be protected by registering the DMFB-specific bioassay pathway to a trusted organization in the same way that benchtop bioassay descriptions are protected.

Watermarking: The bio-protocol parameters are watermarked by embedding a secret signature in the parameter values, whose acceptable range is experimentally determined. The signature is derived by hashing a meaningful message such as the IP name and the owner's identity. This requires a one-way hash function that

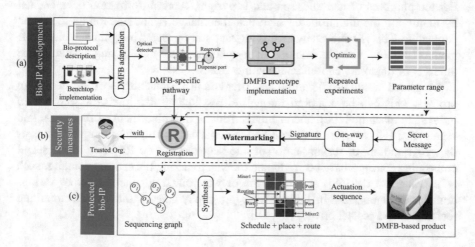

Fig. 5.3 Watermarking a bio-protocol implementation on a DMFB

maps an arbitrary-length message to a fixed-length signature [3]. In the subsequent sections, we describe the embedding of the secret signature in the following bio-protocol parameters:

1. The parameters that are input to the synthesis tool: e.g., mixing time, incubation time, and reagent volume
2. The parameters that are input to the control path: e.g., a valid range of sensor values and error threshold
3. Reagent's mixing ratio in the sample preparation step

To verify the ownership of IP, the developer demonstrates the parameter range and derivation of the parameter values using the hash. Alternately, the parameter range can be stored with the trusted organization along with the pathway registration (Fig. 5.3). The developer can also quantify the ownership claim, as described in Sect. 5.6.

5.3 Related Prior Works

In this section, we describe the related work in other domains and highlight the key differences that are inherent in the proposed watermarking of a bio-protocol.

5.3.1 Watermarking in Various Domains

IP designs across multiple domains have successfully adopted watermarking to deter IP theft [50, 106]. Digital-media refer to audio, image, video, and text data [49]. For the purposes of identification and copyright, secret information is embedded by modifying the digital-media such that the change is not susceptible to human perception. This is achieved by modifying randomly selected insignificant bits of the media [129]. For example, images can be watermarked by modifying the luminance intensity of randomly chosen image points [62].

Integrated circuits (IC) can be watermarked at various stages of the IC design process. This can be achieved through either adding additional transitions in a finite state machine [106], and/or unique routing and placement of functional and scan cells [44]. Software code can be watermarked by inserting unique strings or the programming of responses to specific sequences [48]. Recently, these secret embedding concepts have been adopted for watermarking IPs in other domains, such as neural networks [68]. A comparison of watermarking schemes across various domains is summarized in Table 5.1. Cryptography is used to assist watermarking techniques and protect privacy [34, 58].

Table 5.1 Watermarking in Various Domains

Domain	Digital-media	IC	Neural network	Bio-protocol
Secret put in	Insignificant data points	Unspecified design space	Signed inputs to model	Parameter choice
Fidelity	Exploits human imperfections	Effects side-channel	Effects model accuracy	Exploits variability
Efficacy relies on	Retention of the watermark with signal processing	Plausibility of computing all IC responses	Drop in training accuracy	Plausibility of determining parameter range

5.3.2 Key Difference in the Proposed Solution

The behavior of an IC is determined by the dynamic responses to a sequence of inputs. Some of the input sequences may be unspecified, or the outputs to specific inputs may be *don't-care*. An IC can be watermarked by a secret-message-based response to the unspecified inputs [52]. On the other hand, a bio-protocol IP is designed for limited input–output responses; hence, it lacks *don't-cares* in its responses. For example, a glucose bio-protocol takes serum as a valid input and provides the glucose concentration within the calibrated range [32]. The low cost of biochips prohibits the design of specific responses in case of unspecified inputs such as water instead of serum.

A digital-media watermark embeds secret information by leveraging the imperfection of human audiovisual systems by preserving subjective fidelity [50]. On the other hand, the bio-protocol output error tolerance is dependent on the application, which is limited by the variability in the bio-protocol. In Sect. 5.1, we show that the bio-protocol parameters have inherent variability, which is not known a priori but *experimentally* derived. We leverage the inherent variability in domain-specific parameters such as mixing ratio, sensor calibration, and incubation time of a bio-protocol for watermarking.

5.4 Watermarking of Bio-protocol Parameters

The bio-protocol parameters, given as inputs to the synthesis tool, are referred to as synthesis parameters. On the other hand, parameters that are input to the control path are referred to as control path parameters. In this section, we describe the watermarking of the synthesis and control path parameters. Table 5.2 summarizes the notations used in watermarking. Further, we apply the proposed technique to a case study of immunoassay bio-protocol.

Table 5.2 Notations used in parameter watermarking

Notation	Meaning
p^i	ith parameter value
$[v_{min}^i, v_{max}^i]$	ith parameter value min-max range
c^i	ith parameter resolution
N_{val}^i	The number of possible discrete values of ith parameter
ω^i	Binary signature for ith parameter
l	The number of bits in binary signature ω^i
QA_i	Boolean quality assessment of ith sensor data
s^i	ith sensor data
E_{th}	Error threshold parameter

5.4.1   Watermarking of Synthesis Parameters

An acceptable range of bio-protocol parameter values is derived through several experiments on the biochip. Based on the signature, the developer chooses the parameter value from the experimentally derived range. Let $p^i \in [v_{min}^i, v_{max}^i]$ be the value of the ith parameter with an acceptable range determined by v_{min}^i and v_{max}^i. Assuming the parameter resolution c^i, the total number of possible discrete real values (N_{val}^i) that p^i can take is

$$N_{val}^i = \frac{(v_{max}^i - v_{min}^i)}{c^i} \tag{5.1}$$

All possible values of the parameter can be encoded by a binary string of length $\lceil \log_2(N_{val}^i) \rceil$. The ceiling function is used to ensure that each possible parameter value is mapped to at least one binary representation. Let ω^i be a secret l-bit binary signature, where $l \geq \lceil \log_2(N_{val}^i) \rceil$. Using ω^i, we embed a watermark in the parameter value by the following formula:

$$p^i = v_{min}^i + \left(\frac{\text{int}(\omega^i)}{2^l} \cdot \left(v_{max}^i - v_{min}^i \right) \right) \tag{5.2}$$

where the function $\text{int}(\omega^i)$ converts the binary signature ω^i into its unsigned integer representation. The parameter value p^i computed by Eq. (5.2) can exceed the resolution limit; to address this, the parameter value is corrected to the nearest value with the correct resolution as below:

$$\tilde{p}^i = \left\lceil \frac{p^i}{c^i} \right\rceil \cdot c^i \tag{5.3}$$

Example 16 An immunoassay protocol in [99] reports an incubation time t^1 in the range [360, 600] sec on a biochip running at 1 Hz clock. Hence, $c^1 = 1$ sec and there are $N_{val}^1 = \frac{(600-360)}{(1)} = 240$ valid possibilities for the incubation time. The developer based on a signature of size $\lceil \log_2(N_{val}^1) \rceil = 8$ bits chooses the parameter value accordingly. Let $\omega^1 = $ '00101110'; then the watermarked incubation time t^1 is 403.12 s. After resolution correction, we have $\tilde{t}^1 = 404$ s.

5.4.2 Watermarking of the Control Path Parameters

The control path design provides an efficient monitoring and correction mechanism to overcome the inherent variability in the bio-protocol implementation on a biochip (Figs. 5.1 and 5.2). We watermark the control path design parameters at the global and local levels.

5.4.2.1 Global Control Path

Bio-protocol global control flow can be modeled as a finite state machine (FSM), as shown in Fig. 5.2. The global control flow decides the course of the bio-protocol based on sensor data at the output of various sub-protocols. Let the acceptable range of sensor data s^i be denoted by the closed interval $[v_{min}^i, v_{max}^i]$, where $v_{min}^i (v_{max}^i)$ is the minimum (maximum) acceptable value of the sensor data. We define a boolean quantity called *quality assessment* (QA_i) of s^i, as follows:

$$QA_i = \begin{cases} good & \text{if } v_{min}^i \leq s^i \leq v_{max}^i \\ bad & \text{otherwise} \end{cases} \tag{5.4}$$

If the sensor reading s^i is in the specified range, i.e., $QA_i = good$, then the bio-protocol proceeds to the next sub-protocol, else if the quality is not satisfactory, i.e., $QA_i = bad$, then the relevant steps are repeated, as shown in Fig. 5.2. We embed a watermark in the acceptable range of sensor values, which are determined experimentally. If the sensor data resolution is c^i, the total number of acceptable sensor values s^i is $N_{val}^i = (v_{max}^i - v_{min}^i)/(c^i)$. All the possible valid sensor values can be encoded by a binary string of length $\lceil \log_2(N_{val}^i) \rceil$. The designer generates two w-bit binary strings ($|w| \leq \lfloor \frac{\lceil \log_2(N_{val}^i) \rceil - 1}{2} \rfloor$) ω_{max}^i and ω_{min}^i. The floor function is used to ensure that ω_{max}^i and ω_{min}^i each can be used to encode not more than half of the valid range. We insert the watermark by modifying the range of acceptable sensor values $[v_{min}^i, v_{max}^i]$, as given by

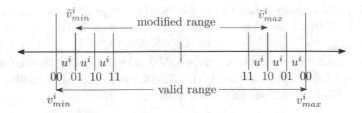

Fig. 5.4 Watermarking of the acceptable sensor value range

$$\tilde{v}^i_{max} = v^i_{max} - (\text{int}(\omega^i_{max}) \cdot c^i)$$
$$\tilde{v}^i_{min} = v^i_{min} + (\text{int}(\omega^i_{min}) \cdot c^i)$$
(5.5)

$\text{int}(\omega^i_*)$ converts the binary signature ω^i_* into its unsigned integer representation. Note that after embedding the watermark, the acceptable range of sensor values becomes $[\tilde{v}^i_{min}, \tilde{v}^i_{max}]$. The output quality assessment is also updated accordingly.

Example 17 The control flow for the quantitative analysis of gene expression is shown in Fig. 5.2 [74]. One of the decision points in the flow is nucleic acid (NA) isolation. A spectrophotometer is used to assess the purity of the isolation of NA and protein. The ratio of observance in the range of 1.68–2.0 is considered to be acceptable. The ratio is calculated to an accuracy of two decimal places, which implies that the resolution is $c^2 = 0.01$. There are 32 valid ratios of observance, and these can be encoded by a 5-bit binary string. Figure 5.4 shows the watermark embedding into the sensor value range. For $\omega^2_{max} = 10$ and $\omega^2_{min} = 01$, we have from Eq. (5.5): $\tilde{v}^2_{max} = 2.0 - 2 \cdot 0.01 = 1.98$ and $\tilde{v}^2_{min} = 1.68 + 1 \cdot 0.01 = 1.69$. Hence, the modified range becomes [1.69, 1, 98].

5.4.2.2 Local Control Path

The local control path monitors the droplet size at intermediate stages. A parameter called *error threshold* E_{th} captures the variance in droplet size due to cutting and evaporation [133]. This parameter can be watermarked similarly to the synthesis parameter (Eq. (5.2)).

5.4.3 Case Study: Particle-Based Immunoassay

Benchtop Protocol Description The immunoassay protocol relies on the antigen–antibody interaction. Antibodies are proteins, generated in response to the invasion of an antigen into the body. Antibodies will bind only to a specific antigen. A benchtop immunoassay protocol performs a sequence of incubation and washing

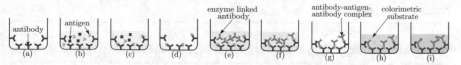

Fig. 5.5 A benchtop immunoassay protocol: (**a**) bottom of the plate is coated with an antibody that binds with a specific antigen of interest, (**b**) clinical sample (urine, serum) is added, (**c**) incubation for antigen–antibody reaction, (**d**) wash, (**e**) enzyme-linked antibody solution is added, (**f**) incubate, (**g**) wash, (**h**) colorimetric substrate is added, and (**i**) chemical reaction between the colorimetric substrate and enzyme-linked antibody–antigen–antibody molecules for colorimetric change

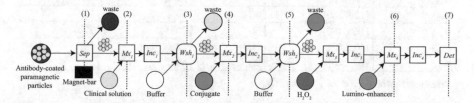

Fig. 5.6 A DMFB implementation for particle-based immunoassay. Mx_i, Inc_i, Det are the mix, incubate, wash, and detection operations, respectively. Sep is the operation for separation of particles from diluent by magnets. Wsh_i is the wash protocol

steps, as shown in Fig. 5.5a–i. A specific antibody is attached to a solid-phase surface (microliter well) (Fig. 5.5a). The test specimen is added and allowed to incubate so that the antigen in the test specimen bind to the antibody (Fig. 5.5b–c). After washing off the non-binding antigens (Fig. 5.5d), a secondary enzyme-linked antibody is added, which binds to the specific antigen (Fig. 5.5e). After that, it is incubated (Fig. 5.5f), and a washing step (Fig. 5.5g) is followed subsequently. Next, a chromogenic substrate is added; the enzyme converts the substrate into a detectable colored compound (Fig. 5.5h–i). The color that develops is proportional to the amount of antigens in the test specimen [30].

DMFB Adaptation In order to reduce the assay time and reagent consumption, the immunoassay is adapted for DMFB implementation. As described in Example 14, the adaptation can be achieved in multiple ways. We describe here a paramagnetic bead-based DMFB adaptation of the immunoassay bio-protocol. In the prototype implementation reported in [99], the DMFB-implementable pathway includes the use of the antibody-coated paramagnetic particles and the external magnetic field. The sequencing graph of the microparticle-based immunoassay is shown in Fig. 5.6. The protocol can be described using the seven steps listed below:

1. A droplet encapsulating antibody-attached paramagnetic particle is dispensed. The particles are separated from the diluent using a magnet.
2. A clinical solution that may contain the antigen/protein of interest is dispensed and mixed with the particles for t_{Mx1} sec and incubated for t_{Inc1} sec.

Table 5.3 Watermarking of synthesis parameters

Parameter name	Valid range $[v_{min}^i, v_{max}^i]$ (sec)	# Values N_{val}^i	Watermark ω^i (hexadecimal)	Parameter value p^i (sec)
t_{Mx1}	[180, 540]	360	33b0	253
t_{Inc1}	[180, 540]	360	15b1	211
t_{Wsh1}	[60, 120]	60	676d	85
t_{Mx2}	[120, 180]	60	aa68	160
t_{Inc2}	[120, 180]	60	16dd	126
t_{Wsh2}	[60, 120]	60	b94a	104
t_{Mx2}	[120, 180]	60	30f4	132
t_{Inc3}	[120, 180]	60	2035	128
t_{Mx4}	[40, 240]	200	7eef	140
t_{Inc4}	[10, 100]	90	4cd1	38

Parameter resolution: $c^i = 1$ sec for all i.

3. The particles are washed in a wash buffer for t_{Wsh1} sec to separate any unbounded protein/antigen.
4. One droplet of the conjugate solution containing enzyme-linked antibody is dispensed, mixed with the particles for t_{Mx2} sec, and incubated for t_{Inc2} sec.
5. The particles are washed in a wash buffer for t_{Wsh2} sec. After the wash, particles are kept suspended in wash buffer and queued for analysis.
6. The particles are separated from the wash buffer and mixed with one droplet of H_2O_2 for t_{Mx3} sec and incubated for t_{Inc3} sec. This droplet is mixed with one droplet of the luminol-enhancer solution for t_{Mx4} sec.
7. The pooled droplet is incubated for t_{Inc4} sec, and the chemiluminescent signal is recorded.

Watermarking We showcase the implementation of the parameter watermarking technique on the described immunoassay. We choose a meaningful message—"New York University Bio-protocol IP" and compute its SHA3-256 hash [3]. The first 160 bits of the resulting hash are used for watermarking various parameters, as shown in Table 5.3. The hash length determines its preimage resistance. A 160-bit hash requires a work of 2^{160} iterations to compute its preimage, which is the same as the minimum preimage resistance recommended in practice [8]. Further, the number of hash bits used for watermarking can be increased by embedding them in various other design steps such as sample preparation [37, 126] or control path design [84, 140].

Recall that to overcome the uncertainties arising from the imperfections of the biochip, feedback control is required. This requires control path design at global and local levels. The immunoassay bio-protocol is divided into five sub-protocols: $Prep_1$, Wsh_1, $Prep_2$, Wsh_2, and $Prep_3$, as shown in Fig. 5.7.

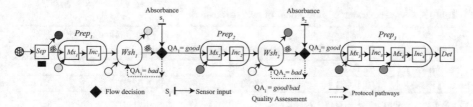

Fig. 5.7 A control path design for immunoassay protocol: $Prep_1$, Wsh_1, $Prep_2$, Wsh_2, and $Prep_3$. Global control flow decision based on the quality assessment (QA_i) of washed droplet based on the sensor input s_i

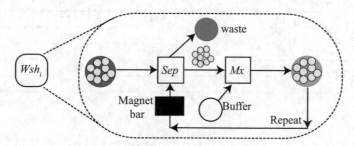

Fig. 5.8 A washing protocol Wsh_i implemented on the DMFB [99], where Sep is the operation for the separation of particles from diluent by magnets, and Mx is the mix operation

The global control path monitors the output of the washing sub-protocol (Fig. 5.7), as it is a critical step in determining the bio-protocol speed [118]. Each wash cycle consists of separating the paramagnetic particles and mixing them with a buffer, as shown in Fig. 5.8. To wash the unbound reagents from magnetic particles to a satisfactory level, the wash cycle is repeated multiple times [46, 99]. This process is verified by measuring the relative absorbance at 450 nm wavelength, as shown in Fig. 5.7. Let s^i be the relative absorbance value measured after Wsh_i wash operation, and then Wsh_i is considered as satisfactory if the sensor value s^i is the range [0, 0.02], where $c^i = 0.001$ [18]. We watermark the upper limit of the valid sensor range using Eq. (5.5):

$$\tilde{v}^i_{max} = v^i_{max} - \left(\omega^i_{max} \cdot c^i \right)$$

The lower limit is not watermarked as 0 absorbance is a trivial lower limit for a wash operation. We use the next 6-bits of the hash value previously computed, as shown in Table 5.4.

The local control path inserts checkpoints within each sub-protocol to monitor intermediate droplet size. A checkpoint is inserted in the control path when the estimated worst-case error limit (E_{lim}) exceeds the error threshold E^i_{th}. We watermark the error threshold E^i_{th} using the equation:

Table 5.4 Watermarking of global control path parameters

Sensor name	Valid range $[v_{min}^i, v_{max}^i]$	Watermark ω_{max}^i	Watermarked range $[\tilde{v}_{min}^i, \tilde{v}_{max}^i]$
s^1 (Wsh_1)	[0, 0.02]	101	[0, 0.015]
s^2 (Wsh_2)	[0, 0.02]	000	[0, 0.02]

Sensor resolution: $c^i = 0.001$ for all i.

Table 5.5 Watermarking of local control path parameters

Sub-protocol name	Valid range (%) $[v_{min}^i, v_{max}^i]$	Watermark ω^i	Watermarked threshold E_{th} (%)
$Prep_1$	[15, 30]	110000	26.25
Wsh_1	[15, 30]	001110	18.28
$Prep_2$	[15, 30]	100010	22.97
Wsh_2	[15, 30]	100111	24.14
$Prep_3$	[15, 30]	010000	18.75

$$E_{th}^i = v_{min}^i + \left(\frac{\text{int}(\omega^i)}{2^l} \cdot \left(v_{max}^i - v_{min}^i \right) \right) \tag{5.6}$$

We use the next 30 bits to watermark the control path in the five sub-protocols of the immunoassay, as shown in Table 5.5.

Therefore, for watermarking the multiple design abstractions (synthesis and control) of immunoassay bio-protocol implementation, we used a hash length of 196 bits. This has a preimage resistance of 2^{196} iterations.

5.5 Watermarking of Sample Preparation

Sample preparation is one of the key steps in most biochemical protocols. It is the process of mixing two or more input reagents in a desired volumetric ratio through a sequence of mixing steps. Several sample preparation algorithms exist in the literature that generates different mixing trees/graphs for the same input ratio [37, 39, 126]. During sample preparation, we embed a watermark in the mixing ratio. This way, the watermark is preserved over different mixing trees/graphs that can be generated using different sample preparation algorithms. Table 5.6 summarizes the notations used in the watermarking of sample preparation.

Table 5.6 Notations used in the watermarking of mixing ratio in the sample preparation step of a bioassay

Notation	Meaning
$\mathcal{M}$	Mixing ratio
R_i	ith input reagent
c_i	Concentration factor (CF) of ith reagent
ϵ	Desired accuracy
$\frac{x_i}{2^d}$	CF of ith reagent in the transformed ratio
b_j^i	jth bit in the binary representation of ith reagent

5.5.1 Basics of Sample Preparation

Formally, a mixture of k input reagents $R_1, R_2, \cdots, R_k$ is denoted as $\mathcal{M} = \{\langle R_1, c_1\rangle, \langle R_2, c_2\rangle, \cdots, \langle R_k, c_k\rangle\}$, where $\sum_{i=1}^{k} c_i = 1$ and $0 \leq c_i \leq 1$ for $i = 1, 2, \cdots, k$, i.e., $R_1, R_2, \cdots, R_i, \cdots, R_k$ are mixed with a ratio of $\{c_1 : c_2 : \cdots : c_i : \cdots : c_k\}$, where c_i denotes the concentration factor (CF) of R_i. Note that $\sum_{i=1}^{k} c_i = 1$ ensures the validity of a mixing ratio. The CF of pure input reagent R_i and a neutral buffer are assumed to be 1 and 0, respectively.

In a (1:1) mixing primitive, which is commonly supported in a DMFB, two unit volume droplets are mixed, and the resulting droplet is split into two equal volume droplets. In the on-chip implementation of sample preparation, the desired target ratio $\mathcal{M}$ is transformed into another ratio that is achievable using the mixing primitive supported by the underlying microfluidic platform and a user-defined accuracy requirement $0 \leq \epsilon < 1$. The transformed mixing ratio is used as an input to the sample preparation algorithms for obtaining a mixing tree or graph representing a sequence of mix–split operations. Figure 5.9 summarizes the sample preparation process. In the case of a DMFB, supporting the (1:1) mixing primitive, each c_i in the input mixture $\mathcal{M}$ needs to be transformed as $\frac{x_i}{2^d}$, by choosing $d \in \mathbb{N}$ appropriately, such that $\max_i\{|c_i - \frac{x_i}{2^d}|\} \leq \epsilon$ and $\sum_{i=1}^{k} \frac{x_i}{2^d} = 1$. Note that d is related to the number of mix–split operations on the mixing tree/graph. The following example illustrates the ratio transformation procedure.

Example 18 Let us assume that we need to generate a target ratio of three input reagents $\{R_1 : R_2 : R_3 = 0.3 : 0.3 : 0.4\}$ $(0.3 + 0.3 + 0.4 = 1)$ on a DMFB supporting the (1:1) mixing model. Let the user-defined error tolerance limit be $\epsilon = 0.001$. By choosing $d = 8$, the target ratio can be transformed as $\{R_1 : R_2 : R_3 = \frac{77}{2^8} : \frac{77}{2^8} : \frac{102}{2^8}\}$ or $\{R_1 : R_2 : R_3 = 77 : 77 : 102\}$. The transformed ratio satisfies the error tolerance limit, i.e., $\max\{|0.3 - \frac{77}{2^8}|, |0.3 - \frac{77}{2^8}|, |0.4 - \frac{102}{2^8}|\} = 0.001 (= \epsilon)$ and $\frac{77}{2^8} + \frac{77}{2^8} + \frac{102}{2^8} = 1$.

For an input ratio $\{R_1 : R_2 : \cdots : R_k = \frac{x_1}{2^d} : \frac{x_2}{2^d} : \cdots : \frac{x_k}{2^d}\}$, a sample preparation algorithm MinMix [126] represents each x_i as a d-bit binary number. Next, these k d-bit binary numbers are scanned from right to left to construct a mixing tree in a bottom-up fashion. Corresponding to each non-zero bit in the d-bit binary representation of x_i, a droplet of input reagent R_i is used as a leaf

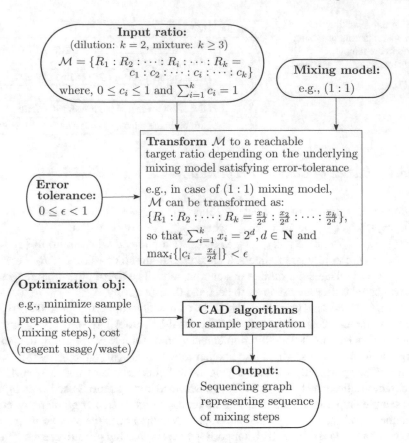

Fig. 5.9 Overview of the sample preparation steps on a DMFB

Fig. 5.10 Mixing tree for the target ratio $\{R_1 : R_2 : R_3 = 77 : 77 : 102\}$ generated by the MinMix [126] algorithm.

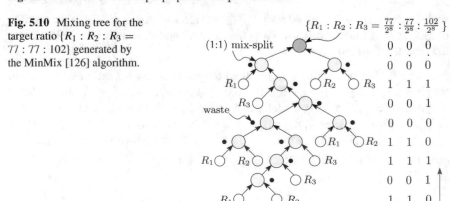

node in the mixing tree. Figure 5.10 shows the mixing tree for the target ratio $\{R_1 : R_2 : R_3 = 77 : 77 : 102\}$ generated by the MinMix.

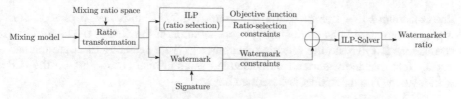

Fig. 5.11 Constraint-based watermarking of mixing ratios

5.5.2 *Constraint-Based Watermarking of Mixing Ratio*

In bio-protocols, several mixtures are required, and the *CF*s of each reagent in the mixture may vary within an allowable range [37]. A bioassay designer may need to perform several experiments for finding an acceptable range of *CF*s that optimizes the assay outcome. We exploit the variability of *CF*s in a mixing ratio for embedding watermark bits. An integer linear programming (ILP) technique was proposed to select a good ratio from the mixture description, where *CF*s of input reagents vary within an allowable range [37]. We impose additional constraints on the ILP formulation to embed a watermark in the selected mixing ratio. Figure 5.11 shows the outline of the proposed watermarking of the mixing ratio.

Assume that *CF*s for input reagents R_1, R_2, and R_3 lie within the concentration range $[0.25, 0.35]$, $[0.27, 0.33]$, and $[0.3, 0.45]$ respectively. Hence, $\{R_1 : R_2 : R_3 = c_1 : c_2 : c_3\}$, where $0.25 \leq c_1 \leq 0.35$, $0.27 \leq c_2 \leq 0.33$, and $0.3 \leq c_3 \leq 0.45$. As seen earlier (Example 18), the error tolerance limit ϵ is set to 0.001, i.e., d is selected as 8 in the ratio transformation. The transformed *CF*s, or R_1, R_2, and R_3, lie within the range $[\lfloor \frac{0.25 \cdot 2^8}{2^8} \rfloor, \lfloor \frac{0.35 \cdot 2^8}{2^8} \rfloor] = [\frac{64}{2^8}, \frac{90}{2^8}]$, $[\frac{69}{2^8}, \frac{84}{2^8}]$, and $[\frac{77}{2^8}, \frac{115}{2^8}]$, respectively. Note that a valid ratio $\{R_1 : R_2 : R_3 = \frac{x_1}{2^8} : \frac{x_2}{2^8} : \frac{x_3}{2^8}\}$ must satisfy $64 \leq x_1 \leq 90$, $69 \leq x_2 \leq 84$, and $77 \leq x_3 \leq 115$, and $x_1 + x_2 + x_3 = 2^8$. There are 396 possible ratios that satisfy the above requirements.

In the proposed watermarking of a mixing ratio, we embed watermark bits into the binary representation of the *CF* of each reagent in the mixing ratio. The pseudo-code of the implementation of this process is shown in Algorithm 1. Suppose the watermark bits are "$\underline{10}\mathbf{0}100111....$" As the numerator of the *CF* of R_1 lies between 64 and 90, i.e., $64 \leq x_1 \leq 90$, we use the least significant $\lceil \log_2(90-64) \rceil$ bits, i.e., 5 bits of the binary representation of x_1 (the numerator of R_1) to embed the watermark. We use $\lfloor \log_2(5) \rfloor = 2$ bits of the watermark (underlined) to select the bit hiding position in the binary representation of x_1 and enforce the next bit (highlighted) of the watermark on the selected bit of x_1. Let the binary representation of x_1 be $(b_7^1 b_6^1 \cdots b_1^1 b_0^1)_2$, i.e., x_1 is represented in 8-bit binary form ($d = 8$). The first two bits (from left) of the watermark are "10." We enforce the third watermark bit, i.e., bit-0, at position two of the binary representation of x_1; hence, $b_2^1 = 0$. We use next $\lfloor \log_2(\lceil \log_2(84-69) \rceil) \rfloor = 2$ watermark bits, i.e., "10," for selecting the bit position in the binary representation of $x_2((b_7^2 b_6^2 \cdots b_1^2 b_0^2)_2)$, and enforce $b_2^2 = 0$. Similarly, the constraint $b_3^3 = 1$ enforces the watermark bits "111" into the CF of R_3. After fixing watermarked bits on the binary representation of the CFs in the target ratio, we use the ILP-based ratio selection technique [37] for selecting the watermarked ratio. Note that the watermark is embedded as additional constraints on the ILP formulation. The ILP model is described below.

$$\text{Minimize: } \sum_{i=1}^{3} \sum_{j=0}^{7} b_j^i$$

Subject to:

1. Ratio selection constraints [37]:

$$64 \le \sum_{j=0}^{7} 2^j \cdot b_j^1 \le 90 \qquad\qquad 69 \le \sum_{j=0}^{7} 2^j \cdot b_j^2 \le 84$$

$$77 \le \sum_{j=0}^{7} 2^j \cdot b_j^3 \le 115 \qquad\qquad \sum_{i=1}^{3} \sum_{j=0}^{7} 2^j \cdot b_j^i = 2^8$$

2. Watermark constraints:

$$b_2^1 = 0 \quad b_2^2 = 0 \quad b_3^3 = 1$$

If we use an ILP-solver [14] without enforcing watermark constraints, the optimized target ratio is $\{R_1 : R_2 : R_3 = 80 : 80 : 96\}$ and the corresponding mixing tree is shown in Fig. 5.12a. However, if we enforce watermark constraints, we get the watermarked ratio $\{R_1 : R_2 : R_3 = 72 : 80 : 104\}$ (Fig. 5.12(b) shows the watermarked mixing tree). Note that the number of mix–split steps in the watermarked mixing tree is increased by one.

Note that 59 different possible mixing ratios that can have a similar watermark. The likelihood of someone else generating such a solution by chance is $59/396 = 0.14$. However, there are only two different solutions that optimize the desired objective function. Hence, if we consider the mixing ratios that optimize the objective function, the likelihood value decreases to $2/396 = 0.005$, a significant decrease in the likelihood of a successful guess.

So far, we have used the inherent variability in the CFs of reagents for inserting watermark bits in the mixing ratio. There are other ways to embed a watermark in the mixing ratio. For instance, we can also exploit the error tolerance limit (ϵ) for watermarking. In sample preparation, we transform the target ratio (by choosing d that determines the number of bits) into a reachable ratio depending on the mixing model and error tolerance (Example 18). We may use a larger value of d, i.e., decreasing accuracy level than specified, for inserting watermark bits. Note that

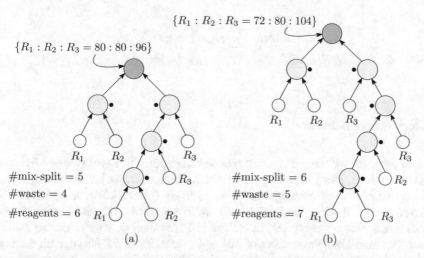

Fig. 5.12 Mixing tree (**a**) non-watermarked and (**b**) watermarked

---

Algorithm 1 watermarkMixingRatio

Input: W: signature, Mixture: $\mathcal{M}$ = $\{\langle R_1, [c_{1,min}, c_{1,max}]\rangle,$ $\langle R_2, [c_{2,min}, c_{2,max}]\rangle, \cdots, \langle R_k, [c_{k,min}, c_{k,max}]\rangle\}$, ϵ: accuracy
Output: Watermarked mixing ratio: $\{R_1 : R_2 : \cdots : R_k = x_1 : x_2 : \cdots : x_k\}$
Set the value of d depending on the ϵ; /* Refer Example 18 */
Approximate $\mathcal{M}$ as $\mathcal{M}' = \{\langle R_1, [\frac{x_{1,min}}{2^d}, \frac{x_{1,max}}{2^d}]\rangle, \langle R_2, [\frac{x_{2,min}}{2^d}, \frac{x_{2,max}}{2^d}]\rangle, \cdots, \langle R_k, [\frac{x_{k,min}}{2^d}, \frac{x_{k,max}}{2^d}]\rangle\}$,
where $x_{i,min} = \lfloor c_{i,min} \cdot 2^d \rfloor$ and $x_{i,min} = \lfloor c_{i,max} \cdot 2^d \rfloor$ for $i = 1, 2, \cdots, k$
/* Watermark insertion */
for $(i = 1; i \leq k; i = i + 1)$ **do**
   $n = \lfloor \log_2(\lceil \log_2(x_{i,max} - x_{i,min}) \rceil) \rfloor$ **if** $(n > 0)$ **then**
      m = readNextWatermarkBits(W, n) /* readNextWatermarkBits(W, n)
      reads next n bits of watermark W */
      $w = $ readNextWatermarkBits$(W, 1)$ Let $x_i = (b^i_{d-1}b^i_{d-2} \cdots b^i_1 b^i_0)_2$ Set $b^i_m = w$;
      /* watermark constraint */
   end
end
Use ILP-based method proposed in [37] for finding the watermarked mixing ratio $R = \{R_1 : R_2 : \cdots : R_k = x_1 : x_2 : \cdots : x_k\}$ /* Watermark constraints are added to the ILP-formulation [37] */
return R

---

increasing the value of d increases the number of bits used to represent each *CF*, and these extra bits can embed the watermark. The watermark can be embedded in the objective function in terms of input reagent costs. In the preceding discussion, we assumed a uniform cost of each reagent, i.e., the cost of R_i $c(R_i) = 1$. We can add a small fractional number (δ_i) to $c(R_i)$ depending on the watermark bits. Here, the objective function needs to be modified, and for our running example, the desired objective function becomes

$$\text{Minimize: } \sum_{i=1}^{3}(c(R_i) + \delta_i) \cdot \sum_{j=1}^{7} b_j^i$$

5.5.3 Case Studies

We experiment with several mixing ratio specifications (five synthetic and five real-life) given in [37] for embedding a watermark. In the watermark embedding process, we derive additional watermark constraints using ("10011001010001110110") as a watermark bit string and solve both ratio selection [37] and the watermark constraints together using an ILP-solver [14] for finding a watermarked mixing ratio that minimizes the number of mix–split steps. Table 5.7 shows both the non-watermarked and watermarked mixing ratios. Note that the increase in the number of mix–split operations in the watermarked mixing ratio determines the watermarking overhead. In Table 5.7, we ignore denominators (2^d, where $d = 8$) of the CF range from each ratio specification.

From the experimental results given in Table 5.7, it can be observed that the embedding capacity, i.e., the number of inserted watermark bits, increases with the increase of the number of input reagents and variability of each CF range in the ratio specification. The probability of coincidence decreases with the increase in embedding capacity. Note that the proposed constraint-based mixing ratio watermark technique cannot embed any arbitrary binary string as a watermark (see row four in Table 5.7). In this case, we change the watermark and repeat the embedding process. In the proposed method, the number of mix–split steps increases as an overhead. The proposed technique guarantees strong authorship with a relatively small increase in the number of mix–split steps in sample preparation.

5.6 Security Analysis

A robust IP protection mechanism protects the IP as a whole and its parts. We achieve this through two steps—*registration* of the design pathway and *watermarking* of parameters. Innovations in pathway design include the use of unique materials like biotin-poly(L-lysine)-graft-poly(ethylene glycol) for antibody binding or developing novel sub-protocols like fluid force discrimination (FFD) for washing [102]. Registering these innovations with a trusted organization protects the rights of the developer, similar to pharmaceutical patents. However, the registration does not cover the entire bio-protocol IP. The developer derives a valid parameter range through repeated trials, and this knowledge is protected through embedding a watermark. The desirable features of the watermark are high credibility for proof

Table 5.7 Embedding of a watermark in the mixing ratio

Mixing ratio space	#Bits	Without watermark Ratio	With watermark Ratio	Δ#mix-split
{[139,179]:[45,60]:[36,46]}	5	160:56:70	162:49:45	3
{[77,102]:[52,76]:[57,71]:[13,38]}	10	96:64:64:32	96:68:64:28	3
{[103,128]:[47,64]:[47,64]: [6,25]:[9,12]}	12	128:64:48:6:10	128:52:48:16:12	1
{[64,76]:[52,76]:[39,51]:[26,51]: [13,25]:[8,20]}	10	64:64:48:48:16:16	No solution found	
{[41,51]:[29,46]:[26,76]:[26,102]: [26,51]:[26,76]:[13,25]}	17	48:32:32:32:32:64:16	49:33:64:36:32:26:16	5
{[47,67]:[18,38]:[1,16]:[1,16]: [1,16]:[1,13]:[140,160]}	13	64:32:8:2:2:4:144	64:32:6:1:8:1:144	1
{[92,112]:[16,36]:[1,13]: [1,13]:[112,132]}	11	96:16:8:8:128	104:17:4:3:128	3
{[16,36]:[5,25]:[41,61]:[16,36]: [1,15]:[1,11]:[117:137]}	18	32:16:48:16:4:4:8:128	32:16:48:16:10:1:1:132	2
{[170,190]:[16,36]:[16,36]:[1,15]: [1,15]:[4,24]}	14	176:32:16:8:8:16	176:32:32:1:2:13	2
{[16,36]:[1,15]:[1,15]:[1,15]:[1,15]: [3,23]:[16,36]:[1,11]:[147,167]}	20	32:1:1:1:4:16:32:8:160	32:1:1:8:4:16:24:2:160	1

100110010100011110110 is used as a watermark

of ownership and resilience to attacks [106]. The following three metrics determine the strength of bio-protocol watermarking:

5.6.1 Probability of Coincidence

The attacker would also stake a claim to the IP. To refute this, the original designer should produce a strong case for authorship. The authorship credibility depends on the probability of coincidence P_c that a non-watermarked implementation carries our watermark. A low value of P_c indicates the strong credibility of authorship. Let p_c^i be the probability of coincidence that the ith watermark constraint is satisfied in a non-watermarked design. Then the probability of coincidence of all watermark constraints is

$$P_c = \prod_{i=1}^{k} p_c^i \tag{5.7}$$

where k is the total number of watermark constraints. The proposed watermark technique involves the choice of various parameters (p^i) from a range of the valid number of choices N_{val}^i. The probability of coincidence for ith watermark constraint is $p_c^i = \frac{1}{N_{val}^i}$. Therefore, from Eq. (5.7) we have

$$P_c = \prod_{i=1}^{k} \frac{1}{N_{val}^i} \tag{5.8}$$

In the watermarking applied to immunoassay bio-protocol, the probability of coincidence is $P_c = 0.93 \times 10^{-20}$, as computed using Eq. (5.8) and N_{val}^i from Table 5.3.

5.6.2 Resilience to Finding Ghosts

An attacker could find a set of constraints that are not part of a watermark but were discovered after the fact. The attacker could then claim it as a watermark. The secret being inserted as a watermark is processed by hash SHA3-256 so that the problem of finding a ghost watermark is at least as difficult as finding a hash preimage [3]. The attacker can group some of the parameters to extract a hash value. However, to attribute this hash to a meaningful message, the attacker needs to compute the preimage of the hash. In the case study of immunoassay, computing the preimage of a hash requires 2^{160} iterations.

5.6.3 Resilience to Tampering

An attacker could try to alter protocol parameters in order to tamper with the watermark. However, the attacker does not know which of the design parameters are used in watermarking to begin with. Without the knowledge of the acceptable parameter range, such tampering of parameters leads to inaccurate results. However, minute changes to the parameter can succeed, if the tampered parameter value lies in the acceptable range. To address this, we define two metrics: ownership score (S_{owner}) and the probability of successful tampering (P_t).

Let the genuine bio-protocol developer watermarked k parameters, where each parameter has N^i_{val} valid number of choices, and then the ownership score (S_{owner}) is given by

$$S_{owner} = \sum_{i=1}^{k} \left(N^i_{val} - \Delta p^i \right) \tag{5.9}$$

where Δp^i is the absolute difference between the observed ith parameter value and the watermarked ith parameter value. In the case of two parties contesting the ownership of the IP, the one with a higher score S_{owner} has a stronger claim. To successfully tamper with the watermarked parameter values, the attacker aims to lower the ownership score S_{owner} while keeping the tampered parameter within the valid range $[v^i_{min}, v^i_{max}]$, i.e., $v^i_{min} < p^i \pm \Delta p^i < v^i_{max}$. Assuming that p^i has a uniform probability over its valid range, the probability that $p^i \pm \Delta p^i$ is within the valid range $[v^i_{min}, v^i_{max}]$ is $\left(1 - \frac{2\Delta p^i}{v^i_{max} - v^i_{min}} \right)$. Considering the tampering of k individual parameter values as an independent event, the probability that all k tampered parameter values are within the valid range is obtained by multiplying the corresponding probabilities of k parameters. Therefore, the probability of successful tampering (P_t) is

$$P_t = \prod_{i=1}^{k} \left(1 - \frac{2\Delta p^i}{v^i_{max} - v^i_{min}} \right) \tag{5.10}$$

The score S_{owner} decreases linearly as the probability P_t decreases exponentially with the extent of tampering, as shown in Fig. 5.13. If the attacker tries to increase Δp^i, probability P_t decreases. On the other hand, if the attacker tries to be safe by choosing small Δp^i, the score S_{owner} is decreased by a small amount. The probability of successful tampering P_t is as low as 0.005, with the score S_{owner} lowered by 31 from 1370 to 1339. This shows that without the knowledge of the valid range, tampering with the watermarked parameter values is very less probable, and the attacker is forced to do rigorous experiments. We demonstrate this through the following attacks:

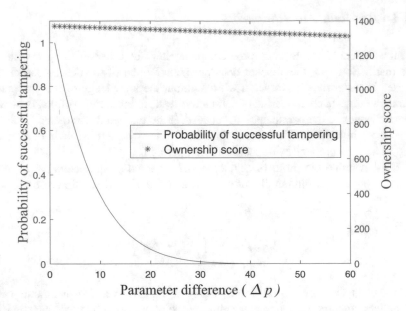

Fig. 5.13 The probability of successful tampering (P_t) reduces exponentially for small tampering with watermarked parameter values $\pm\Delta p$, whereas ownership score (S_{owner}) sees only a linear reduction with $\pm\Delta p$

5.6.3.1   Brute Force

The attacker tries to reverse engineer the parameters through multiple trials of the bio-product and reconstructs the range of each parameter.—This is unlikely to succeed because of the complex nature of the interplay of bio-protocol parameters. A developer invests high effort and money in meticulously analyzing the interplay of these parameters to determine its range. Without the knowledge of the interplay of the parameter, the attacker would have to do a large number of trials. Each of these trials costs in terms of the biochip hardware, reagents, and time. This goes against the economic objective of the attacker to steal bio-protocol IP. The purpose of the watermark is served if the effort required is comparable to the development of the protocol parameters through rigorous research and experimentation.

5.6.3.2   Insert Own Signature

The attacker tries to embed their own signature by guessing the parameter range around the known value.—This is a feasible attack but cannot trump the score S_{owner} of the owner. The ownership score is proportional to the valid range (N_{val}^i) of watermarked parameters. Even if the attacker chooses a small subset of the valid range ($n_{val}^i < N_{val}^i$) and inserts a signature, the resulting score will be based on the subset of the valid range and so will be lower than the owner's score (Eq. (5.9)).

The owner and the attacker can have a similar score only if the attacker knows the parameter range.

5.6.3.3   Watermark More Parameters

The attacker tries to watermark a bigger subset of parameters to increase the ownership score, thereby increasing the number of watermarked parameters k in Eq. (5.9).—This is unlikely to succeed because the attacker does not know which parameters are watermarked to begin with. Further, the developer can easily avoid this attack by watermarking a large number of parameters. This can be done by using a longer hash. Alternatively, the knowledge of parameters chosen for the watermark can be stored with the trusted organization, as discussed in Sect. 5.2. This can deter any wrongful ownership claims by watermarking new parameters.

Chapter 6
Obfuscation for IP Protection

Application Context Consider a bioassay developer who invests heavily in bioassay IP development. Such a developer can be either a pharmaceutical company performing drug trials or developing personalized medicine for its customers. The developers use microfluidic platforms, manufactured in-house, to conduct large-scale (high-throughput, parallel, and automated) experiments [90]. Continuous flow-based microfluidic biochips (CFMBs) have evolved rapidly in the last decades [90, 91]. The CFMBs allow automated control of fluid flow in a network of microchannels by suitable actuation of pressure-driven microvalves [91]. The biochip controller is connected to the network for round-the-clock online monitoring and control [17]. The user can focus the CCD camera on the area of interest to verify the biochip's state. We discuss one such context in the below example.

Example 19 The CFMB showcased in Fig. 6.1 is used for evaluating the response of the particular dose of a drug on the bacterial cells. The drug is loaded through one of the inlets, and a dilution buffer can be loaded through the other inlet. The drug is diluted through a sequence of load and mix operations. The resulting diluted drug is then injected into a chamber of living cells, and its response is recorded. This experiment is repeated for various concentrations to find the minimum drug concentration required to kill the bacterial cells.

The Threat of IP Theft An attacker can be a competitor who is motivated to steal the IP from the developer without incurring any cost of development. Thus, the attacker gains economically by accessing such high-value IP, i.e., the bioassay. To reverse-engineer the bioassay, the attacker accesses the actuation sequence and the snapshots of the biochip layout. The attacker can then map the actuation sequence to fluidic operations and rebuild the bioassay, as shown in Sect. 2.3.1. The attacker

---

Based on "Thwarting Bio-IP Theft Through Dummy-Valve-based Obfuscation", TIFS 2021 [112].

© The Author(s), under exclusive license to Springer Nature Switzerland AG 2022
S. Mohammed et al., *Security of Biochip Cyberphysical Systems*,
https://doi.org/10.1007/978-3-030-93274-9_6

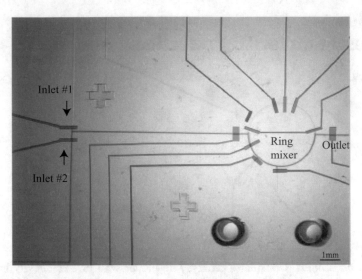

Fig. 6.1 A continuous-flow biochip with two inlets, one ring mixer and one outlet

accesses these through a network attack. A recent ransomware cyberattack in the pharmaceutical industry underlines the seriousness of this threat [10].

The biochip controller is connected to the network for around-the-clock online monitoring and control [17]. Further, the actuation sequence is stored in the biochip controller to support online error correction [74]. A remote attacker can gain access to the network [78, 103]. Then the attacker can launch one of the two attack levels:

1. The attacker uses the network for a short duration to access a single image of the biochip layout and the actuation sequence of the bioassay. It is more likely that the attacker can escape detection for a short duration. This becomes more practical when the bioassay process is not being actively monitored.
2. The attacker logs on to the network through the entire period of bioassay execution to gain access to (high- or low-resolution) images of all stages of the execution as well as the actuation sequence. This increases the chances of detection of unauthorized access.

The attacker can differentiate between the pressurized valve and de-pressurized valve due to the visual difference. The attacker does not have access to the CFMB. However, the attacker can build a prototype from the biochip layout snapshots. Such a prototype can be used to remove any ambiguity left in the reverse-engineering process.

Solution Outline To thwart the reverse-engineering of bioassays, we need to obfuscate the one-to-one mapping between the actuation sequence, biochip layout, and fluidic operations. This can be achieved by careful insertion of dummy-valves in the biochip [93]. In other words, the use of dummy-valves along with normal valves

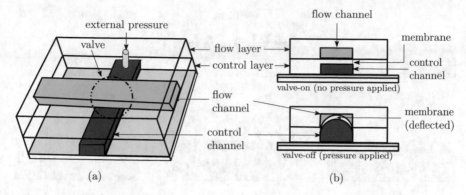

Fig. 6.2 Schematic of a sieve valve: (**a**) top view and (**b**) cross-section view

obfuscates the biochip layout and the actuation sequence. Without the knowledge of the type of the valve (normal/sieve), fluidic operations cannot be determined [93].

6.1 Obfuscation Primitive

Here, we describe the structure of obfuscation primitives—sieve valve and multi-height-valve. Next, we compare the primitives for obfuscated biochip fabrication. We provide simulation and experimental results to establish the efficacy of the multi-height-valve as an obfuscation primitive.

6.1.1 Sieve Valve

In a normal valve, the flow channel is semi-circular shaped. When the valve is pressurized, it seals the flow channel (Fig. 1.3b). However, if the flow channel is rectangular, the pressurized valve membrane partially closes the flow channel, as shown in Fig. 6.2. This is the sieve valve (Fig. 6.2b) [107]. These are used in CFMBs to trap cells. Closing the sieve valve blocks the cells but allows the fluid to pass through [135]. Fabrication of rectangular flow channels requires a negative photoresist instead of a positive photoresist, which is used for normal valve fabrication.

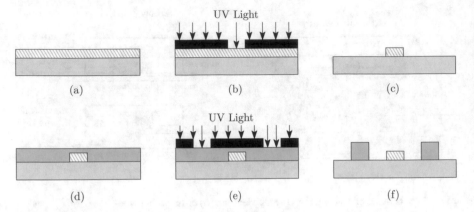

Fig. 6.3 Fabrication of multi-height structure: a cross-section view of (**a**) spin coating with photoresist, (**b**) UV exposure, (**c**) removal of uncured photoresist, (**d**) spin coating with photoresist, (**e**) UV exposure, (**f**) removal

6.1.2   Multi-Height Valve

The amount of pressure needed to open or close a valve is determined by the membrane thickness in the valve region [47]. If the height of the control channel is lowered, then it results in a thicker membrane, as shown in Fig. 6.3. This requires a higher pressure to operate (close/open) compared to the normal membrane. When it is operated at a lower pressure, the valve does not close/open completely [72, 79]. For example, the work in [79] shows that a 34 μm membrane valve requires a minimum pressure of 12 psi to operate, whereas a 28 μm membrane requires a minimum pressure of 8 psi to operate.

6.1.3   Ideal Obfuscation Primitive

An ideal obfuscation primitive would be easy to integrate with the biochip and hard to differentiate from a normal valve. A sieve valve results in a discontinuity in the flow-layer shape from semi-circular (normal) to rectangular. While biochips have only a few sieve valves [135], obfuscation requires the insertion of one extra sieve valve for every normal valve. This leads to a corresponding increase in fabrication complexity. On the other hand, the fabrication of a multi-height-valve requires modification of the height of the control valve. This can be easily achieved by simply controlling the photoresist viscosity or spin speed or number of spin iterations. In other words, multi-height-valves are easier to integrate with a biochip and easier to scale up in terms of their numbers than sieve valves.

Next, we establish the efficacy of the multi-height-valve in acting as an obfuscation primitive. We present experimental evidence through a prototype multi-height-

Fig. 6.4 A multi-height valve as an obfuscation primitive. The normal and dummy valves are both open even though the normal valve is de-pressurized, and the dummy valve is pressurized. As a result, both fluids #1 and #2 are flowing, as shown in the dotted circle

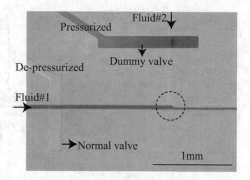

valve fabrication and subsequent demonstration, as shown in Fig. 6.4. Using this, we were able to obfuscate the sample biochip shown in Fig. 6.1. The similar images of the valves provide validation for the efficacy of multi-height valves as an obfuscation primitive. We also provide a link to a video that demonstrates that there is no obvious visible difference between the operation of a dummy and a normal valve.[1] We describe this in more detail in Sect. 6.5.

6.2   Obfuscation for IP Protection

To deter the reverse-engineering of a bioassay, we propose to obfuscate the actuation sequence by carefully inserting sieve or multi-height valves in the biochip. In the rest of the chapter, we refer to a "sieve or multi-height" valve as a "dummy" valve. The bioassay developer keeps the bioassay description and the dummy-valve locations a secret. The developer uses a CAD tool on a trusted offline computer to synthesize the obfuscated actuation sequence. The obfuscated sequence is loaded in the biochip controllers that are used to conduct the high-valued-experiments, as shown in Fig. 6.5. Minor software updates are handled in the biochip controller, and major updates are performed in the trusted offline computer.

Note that the insertion of only dummy valves or only dummy actuations fails to protect the IP. If only dummy valves are added, then the attacker can reverse engineer the IP by using the actuation sequence. If there are no dummy valves but only dummy actuations, then the attacker's problem is to map the real actuations to the biochip valves. Actuations are the sequence of control signals applied for the valves. The biochip controller converts the stored actuations to the electrical signals to the valves. Even with momentary access to the controller, an attacker can easily prune the real actuations from the dummy actuations by observing the electrical signals provided to the valve. To provide a defense against this attack, we need to insert dummy valves in the biochip.

---

[1] https://youtu.be/RS3fRLoQwSQ.

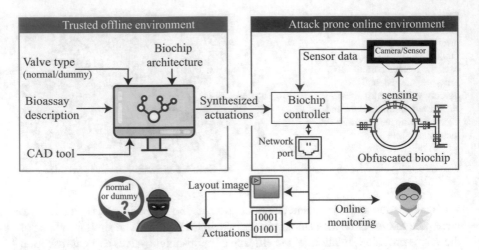

Fig. 6.5 Proposed dummy-valve-based obfuscation technique

Fig. 6.6 Channel $i \rightarrow j$ with
(**a**) valve 1 and (**b**) valves $1a$,
$1b$

Table 6.1 Boolean variables to characterize a channel

Parameter	Description	Interpretation	
		1	0
v_a	Status of a valve	Pressurized	de-pressurized
g_a	Type of a valve	Normal	Dummy
$c_{ij}^k, k \in \mathbb{N}$	Status of channel $i \rightarrow j$	Open	Closed

k: #valves on the fluidic channel connecting ports i and j.

6.2.1 Dummy-Valve-Based Obfuscation

Consider the channel between port i and j, as shown in Fig. 6.6a. Let the channel $i \rightarrow j$ be open if valve 1 is pressurized, else it is closed. Such a valve is a normal valve. On the other hand, if the valve is a dummy, then the channel $i \rightarrow j$ is always open, regardless of the actuation state of valve 1. To capture the differences between a normal and a dummy valve, consider the Boolean variables defined in Table 6.1. Using these variables, we describe the channel in Fig. 6.6a as

$$c_{ij}^1 = \overline{g_1} \vee (g_1 \wedge v_1) = \overline{g_1} \vee v_1 \tag{6.1}$$

Here, "$\wedge$," "$\vee$," and "$\overline{g_1}$" represent Boolean operations "and," "or," and "negation," respectively. As per the attack model, g_1 is secret, and v_1 is known from the actuation sequence. Equation (6.1) captures the obfuscation introduced in the fluid channel characteristics due to the unknown valve type. Without the knowledge of g_1, an attacker does not know the channel status. The dummy valve reduces the flow-rate in the channel. However, this can be neutralized by either increasing the fluid pressure at the inlet or allowing more time for the fluid flow. For the purposes of our analysis, we ignore the change in flow-rate. Consider an increase in the number of valves on the channel, as shown in Fig. 6.6b. The characteristic of the channel is given by following Boolean equation:

$$c_{ij}^2 = (\overline{g_{1a}} \vee v_{1a}) \wedge (\overline{g_{1b}} \vee v_{1b})$$

$$= (\overline{g_{1a}} \wedge \overline{g_{1b}}) \vee (\overline{g_{1a}} \wedge v_{1b}) \vee (\overline{g_{1b}} \wedge v_{1a}) \vee (v_{1a} \wedge v_{1b}) \qquad (6.2)$$

If there are n such valves on a channel $i \rightarrow j$, the characteristic of the channel can be captured as

$$c_{ij}^n = \bigwedge_{\gamma=1}^{n} (\overline{g_\gamma} \vee v_\gamma) \qquad (6.3)$$

Comparing Eqs. (6.2) and (6.3), increasing the number of valves increases the channel obfuscation due to the corresponding increase in the number of unknown parameters (g_*). Using this primitive, we describe the obfuscation of the reagent load operation, the biochip structure, and the bioassay parameters such as mix-time and reagent volume.

6.2.2 Reagent Load Obfuscation

A biochip consists of functional modules such as a fluid inlet/outlet, mixer, storage, reaction chamber, and multiplexer/demultiplexer. As shown in Sect. 2.3.1, the actuation signals of a biochip have a one-to-one mapping to the fluidic operations. We insert dummy valves in the biochip functional modules so that the actuation-signal to fluidic-operation mapping is no longer preserved. Since the valve type is kept secret, the channel characteristic can be obfuscated, as shown in Eq. (6.3). Thus, the attacker cannot determine the fluidic operations correctly to reverse-engineer the sequencing graph (IP). This is called *behavioral obfuscation*.

Consider the biochip shown in Fig. 2.6 with a two input multiplexer and a rotary mixer. It mixes two input reagents R_1 and R_2 in a 3 : 1 ratio, as explained in Sect. 2.3.1. Additional valves (normal and dummy) are added to obfuscate the biochip, as shown in Fig. 6.7a. In the modified CFMB platform, one or more dummy valves on the input to output paths can be de-pressurized to deceive the attacker from identifying the correct fluidic path. From Eq. (6.3), the channel state (open/close) depends on the valve type (g_*), which is unknown to the attacker. The following example illustrates obfuscation on the fluidic path.

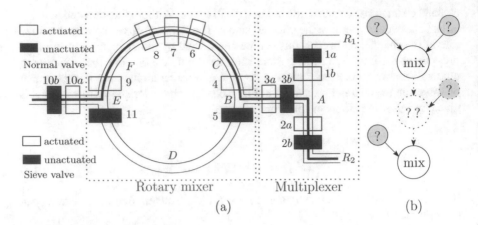

Fig. 6.7 (**a**) Dummy-valve-based obfuscated functional modules of mixer and multiplexer. (**b**) The obfuscated sequencing graph

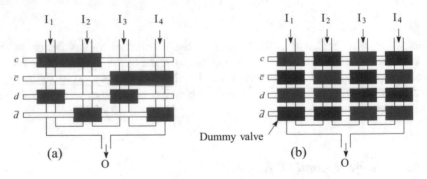

Fig. 6.8 (**a**) A 4-inlet binary multiplexer design. For example, inlet I_1 flows to output "O" when control lines c and d are de-pressurized. (**b**) Obfuscated binary multiplexer

Example 20 In Fig. 6.7a, let $\{1b, 2b, 3b, 10b\}$ be dummy valves and the rest be normal valves, i.e., $g_{1b}, g_{2b}, g_{3b}, g_{10b} = 0000$, then the actuation set $v_{1a}, v_{1b}, v_{2a}, \cdots, v_{10a}, v_{10b}, v_{11} = 011010101111110$ pushes R_2 into the mixer (ref. Fig. 6.7a). On the other hand, if $\{1a, 2a, 3b, 10b\}$ are dummy valves and the rest are normal valves, then the same actuation set will push R_1 into the mixer. Without knowing the valve type (dummy or normal), an attacker cannot determine the inputs to the mix operation in a sequencing graph, as shown in Fig. 6.7b.

As the size of the multiplexer increases, the number of ports increases [91]. A variety of schemes are used to enable the scaling of the multiplexer. A binary multiplexer of N (vertical) inlets requires $2 \log_2(N)$ (horizontal) control channels. Valves are formed only where a wider section of the (horizontal) control channel intersects the vertical flow channel, as shown in Fig. 6.8a. Here, the control channels are binary pairs [91]. In such a case, the multiplexer structure can be easily obfuscated by modifying the thin section of the control channel to a dummy valve, as shown in Fig. 6.8b. The advantage of obfuscating a binary multiplexer is that it does not increase the number of control ports.

6.2.3   Mix-Time Obfuscation

The mixer in Fig. 6.7a has a ring with one inlet channel $A \rightarrow B$ and an outlet channel $E \rightarrow O$. The mixing time can be deduced from a sequence of opening and closing of the inlet/outlet channels followed by the peristaltic pumping operation. The status of the channels $A \rightarrow B$ and $E \rightarrow O$ can be obfuscated by adding extra valves; viz Eq. (6.3). This leads to ambiguity in the mixing time and the number of mixing steps. The following example describes it in detail.

Example 21 In Fig. 6.7a, consider the valve types as in Example 20, i.e., all $g_* = 1$ except $g_{1b}, g_{2b}, g_{3b}, g_{10b} = 0000$. Let the valve actuation be $v_{1a}, v_{1b}, v_{2a}, \cdots, v_{10a}, v_{10b}, v_{11} = 011010101111100$. This opens the mixer inlet/outlet to push out the mixer content, which denotes the end of the previous mixing step. If the actuation is followed by a peristaltic pumping operation, then it denotes the start of a new mixing. On the other hand, if $\{3a, 10a\}$ are dummy valves, and $\{3b, 10b\}$ are normal valves, then the given actuation set does not open the inlet/outlet of the mixer. Hence, the previous mixing step has not ended, and a new mixing step has not started. This leads to obfuscation in the deduction of the mixing time and the number of mixing steps, as shown by the dotted nodes in the sequencing graph in Fig. 6.7b.

6.2.4   Structural Obfuscation

The *behavioral obfuscation* does not change the structure of the biochip but inserts extra valves on the existing channels. Furthermore, the structure of the biochip can be obfuscated by inserting dummy channels, multiplexers, and/or mixers. This is *structural obfuscation*. A channel can be mimicked by a dummy multiplexer with a dummy valve on the original inlet—so that it is always open and a normal valve on a dummy inlet—that is kept closed. Without the knowledge of the valve type, the attacker cannot know which inlet is selected when both the valves are closed. Alternately, the channel can be mimicked by a dummy mixer with dummy valves forming an always open channel in the ring mixer. The valves of this module are pressurized like a mixing module to mislead the attacker. To resolve this ambiguity, an attacker has to do trial and error by replacing each mixing operation in the actuation with a transportation operation.

Example 22 In Fig. 6.9, ports $12b$, $13b$ are dummy valves and ports $12a$, $13a$ are normal valves. For actuation set $v_{12a}, v_{12b}, v_{13a}, v_{13b} = 0000$, paths $R_1^2 \rightarrow R_1$ and

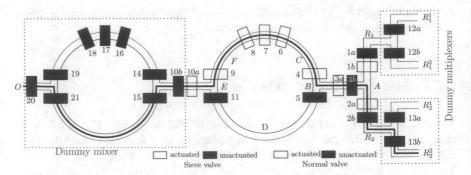

Fig. 6.9 Structural obfuscation: a dummy multiplexer addition at inlets R_1, R_2 and a dummy mixer addition at outlet O

$R_2^2 \rightarrow R_2$ are open. On the other hand, if ports $12b$, $13b$ are normal valves and ports $12a$, $13a$ are dummy valves, then for the same actuation set $R_1^1 \rightarrow R_1$ and $R_2^1 \rightarrow R_2$ are open. This leads to the obfuscation of the fluid selected. Furthermore, a dummy mixer with $\{15, 20, 21\}$ as dummy valves is added to path $E \rightarrow O$. The valves of this mixer can be pressurized to mimic a normal mixer, whereas, in reality, it is a $E \rightarrow O$ channel controlled by valve port $10a$.

6.2.5   Reagent Volume Obfuscation

A bioassay implementation requires the mixing of reagents in measured quantities. The reagent volume is a key parameter that determines a bioassay's outcome and its precision [126]. The bioassay developer finds the reagent volume parameter through numerous trials on the biochip [46]. A metering block can be used to measure different quantities of reagent before loading them, and we elucidate it through the following example.

Example 23 Consider a metering circuit shown in Fig. 6.10a. The regular 1:1 mixing can be achieved by: (1) A fluid can be loaded by opening valves *in1*, *1–4*, *2*, and *o2*. (2) Another fluid can be loaded by opening valves *in1*, *5–8*, and *o2*. Further, 3:1 can be achieved by: (1) A fluid can be loaded by opening valves *in1*, *1–6*, and *o3*. (2) Another fluid can be loaded by opening valves *in1*, *7–8*, and *o3*. In other words, it can support different mixing ratios such as 1:3, 3:1, 1:1 of two input fluids, as shown in Fig. 6.10b–e.

The mixing ratio can be estimated through the actuation states of the metering block valves. This information can be obfuscated using additional dummy valves, as shown in Fig. 6.10f. If the original biochip does not have a metering block, then the biochip can be modified to mimic a metering block to enhance the obfuscation. This is achieved by dividing the mixer into N equal segments using extra N valves—

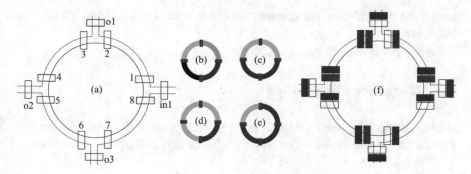

Fig. 6.10 (**a**) A mixer with a metering capability. The mixer can be used to mix fluids in different ratios—(**b**) 1:1:1:1, (**c**) 1:1, (**d**) 3:1, and (**e**) 1:3. (**f**) A mixer with an obfuscated metering block

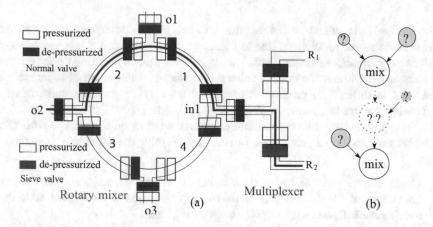

Fig. 6.11 (**a**) Obfuscated mixer with a metering circuit. The peristaltic pumps are excluded for simplicity. (**b**) DAG description of the obfuscated fluid selection and metering volume. The droplet type and volume are unknown

referred to as ratio valves. An input fluid can be filled in 0 to N of these segments. This metering of fluid can be obfuscated by adding dummy valves alongside the ratio valves. Without the knowledge of the dummy and normal valves, an attacker cannot determine the number of segments filled by a fluid.

Example 24 Consider the bioassay described in Sect. 2.3.1. We replace the simple mixer with a metering circuit, as shown in Fig. 6.11a. With this modification, reagents can be loaded in variable volumes (0, 1, 2, 3, or 4 parts). A targeted mixer with $R_1 : R_2 = 3 : 1$ can be achieved by loading R_1 in 3 parts of the meter and R_2 in one part, followed by a mixing operation. On the other hand, using the simple mixer, we require three load and two mix operations, as described in Sect. 2.3.1. In other words, the meter circuit reduces the number of operations. On the other hand, the metering circuit enables obfuscation of reagent volumetric information. Without the knowledge of the valve type, the attacker cannot determine which reagent was loaded and the volume of the reagent. This obfuscation is described by the DAG in Fig. 6.11b.

6.3 Design for Obfuscation

Having outlined the proposed obfuscation technique, we next define the security metrics that capture the security-cost trade-offs and design-for-obfuscation rules.

6.3.1 Undoing the Obfuscation

To reverse-engineer the bioassay, the attacker has to interpret the actuation sequences that are ambiguous due to unknown valve type g_*. Such actuations are referred to as *ambiguous actuations*. The attacker can build a biochip prototype from the snapshots without the correct valve types. Note that the result of biochemical reactions is difficult to predict; developers often rely on experimental trials to determine the results, such as drug trials [90]. Moreover, the adaptation of a benchtop protocol to a biochip requires experimental trials on a prototype [46]. The attacker can use such a prototype as an oracle to resolve the ambiguous actuations by:

1. *Crude attack* By trial and error, the attacker can replace the ambiguous actuations until the results of the bioassay on the biochip prototype become identical to the known results (such as the targeted bacteria is killed).
2. *Knowledge-based attack* The attacker can use some of the information from the benchtop protocol to deduce the bioassay. We discuss this issue in more detail in Sect. 6.4.

The maximum number of experiments required to resolve this ambiguity is referred to as *resolution effort $\mathcal{E}$*. This is used as a metric of the efficacy of an obfuscation method. The design overhead for obfuscation is defined in terms of extra valves, which in turn may lead to extra pins in the biochip and extra memory for storing the corresponding actuation signals. The biochip designer obfuscates the biochip and its actuation sequence to make reverse-engineering hard enough to deter an attacker. To maximize the resolution effort, we propose the following design rules.

6.3.2 Design-for-Obfuscation Rules

In a crude attack, the attacker will try all combinations of g_*. However, a smart attacker will leverage functional properties to prune the search space. To achieve a robust obfuscated design, we frame four design rules.

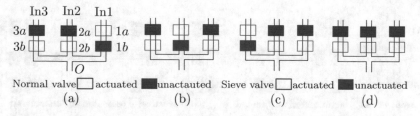

Fig. 6.12 Multiplexer actuation to push fluid through (**a**) In1, (**b**) In2, (**c**) In3, and (**d**) no fluid

6.3.2.1 Channel

A continuous channel needs to be formed from input port to output to push fluid in a CFMB. The attacker tries to identify which input–output path is opened in a given cycle. If there exists an input–output path without any de-pressurized valve, then the actuation is unambiguous to the attacker. Else, the attacker has to guess if any of the de-pressurized valves on the input to output paths is a dummy valve. This leads to the first design rule.

Rule #1: In an ambiguous actuation, every input to output channel path must have at least one closed dummy valve.

Consider a channel that has n_{chl} de-pressurized valves in an ambiguous actuation cycle. Without knowing the valve type g_*, i.e., dummy or normal, the de-obfuscation effort $\mathcal{E}_{chl}$ involves trials that map each closed valve to two possibilities—closed and open. Hence, $\mathcal{E}_{chl} \leq 2^{n_{chl}}$. The effort increases with the number of distinct input–output paths with closed valves in a cycle.

6.3.2.2 Multiplexer

An attacker can use the following properties of a multiplexer to resolve the obfuscation. *P1*: At most, one path of the multiplexer can be open at any time. *P2*: It is likely that each inlet fluid is selected at least once in a bioassay. An attacker can collect all the unique actuations applied to the multiplexer, and along with the properties *P1* and *P2* the attacker can de-obfuscate the multiplexer actuations as discussed in the following example.

Example 25 Consider a 3-inlet multiplexer with two valves $*a$ and $*b$ on each inlet. For each inlet, the set of actuations $v_{*a}, v_{*b} = \{11, 00\}$ is unambiguous, and $v_{*a}, v_{*b} = \{10, 01\}$ is ambiguous. Between any pair of inlets, there are four possible combinations of these ambiguous actuations. In Fig. 6.12, 3-out-of-4 combinations are used for actuating the valves $v_{1a}, v_{1b}, v_{2a}, v_{2b}$. The unused ambiguous actuation combination on the inlet *In1* and *In2* is $v_{1a}, v_{1b}, v_{2a}, v_{2b} = 1010$. The attacker can decipher that this actuation opens both inlets *In1* and *In2* and hence is not used due to property *P1*. Alternately, an attacker can guess that the least used actuation on an inlet is used to open the respective inlet. In Fig. 6.12, actuation $v_{*a}, v_{*b} = 10$ is the

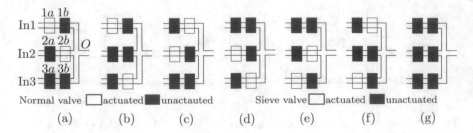

Fig. 6.13 Multiplexer actuation for (**a**)-(**b**) pushing In1, (**c**)-(**d**) pushing In2, (**e**)-(**f**) pushing In3, and (**g**) not pushing any fluid

least used actuation on each inlet. The attacker can decipher with a high probability that these actuations open their respective inlets due to property *P2*.

A naive defense against the above attacks is to increase the number of valves on each inlet. However, this increases cost. To avoid cost escalation, we use two valves per inlet with design rules #2 and #3. Rule #2: Apply ambiguous actuations to no more than two inlets at a time. One inlet is the fluid being pushed and one from the other $m - 1$ inlets of the multiplexer. Rule #3: Apply unambiguous actuations when no fluid is pushed through the multiplexer.

In an m-inlet multiplexer, through these design rules, there are $m - 1$ ways of actuating an obfuscated push operation of a fluid. The ambiguous actuation on the same inlet can be used in the obfuscated push operation of other $m - 1$ inlets. This defeats the two attacks described in Example 25. The maximum number of unique ambiguous actuations is $s_{mux} = \binom{m}{2}$, as shown in Fig. 6.13. Each ambiguous actuation can be mapped to two possibilities. If there are s unique ambiguous operations in a bioassay, an attacker needs to perform 2^s experiments. The maximum resolution effort is $\mathcal{E}_{mux} = 2^{s_{mux}} = 2^{\binom{m}{2}}$.

6.3.2.3 Mixer

Reliable mixing requires a minimum mixing time, which depends on the fluid velocity and channel geometries, etc. If ambiguous actuations are inserted in the mixer actuation sequence prior to the minimum mixing interval, then the attacker can map that actuation to an ongoing mix operation and prune the search space. To avoid this, we frame design rule #4.

Rule #4: The gap between an ambiguous and other mix operations must be more than the minimum mix-time, t_{min}.

The number of ambiguous mixer actuations is dependent on the number of valves on the mixer inlet and outlet, provided rule #4 is satisfied. However, to minimize the cost, we use two valves on the mixer inlet (outlet). The number of possible ambiguous actuations on the mixer inlet (outlet) is two. This implies that the maximum number of ambiguous actuations that can be applied to the mixer (inlet and outlet) is $s_{mix} = 4$. The ambiguous actuations can be mapped to one of the two possibilities—a new-mix operation or no new-mix operation. Therefore, the reverse-engineering effort for a mixer is $\mathcal{E}_{mix} = 2^{s_{mix}} = 2^4$.

6.3.2.4 Dummy Structures

The same rules apply to dummy structures such as multiplexers and mixers (Fig. 6.9). To resolve the ambiguity about n_{dum} dummy structures, ($\mathcal{E}_{dum} = 2^{n_{dum}}$) trial experiments must be performed. However, the cost of introducing dummy structures includes not only extra valves but also extra channels and extra input/output ports.

6.3.2.5 Metering Circuits

Consider a metering circuit that loads a fluid in 1 to n_p parts. Such a metering circuit can be obfuscated by adding a dummy valve along with each original valve, i.e., the total number of valves is doubled. In the case of a n_p-part metering mixer, the number of extra valves is $3n_p$. In order to resolve the ambiguity, an attacker needs to determine which of the outlets are open. Since there are n_p outlets, the reverse-engineering effort is $\mathcal{E}_{meter} = 2^{n_p}$. In the case that the original mixer supports only $1 : 1$ mixing, it has six valves to load the upper and lower half, then the extra valves required to build a n_p-part metering mixer is $6n_p - 6$. All the mixer outlets can be merged into a single biochip waste outlet.

6.4 Special Cases

The output of a bioassay is the result of the interplay of multiple factors—the biochemistry of the reagents and parameters such as volume, mix, and incubation time. This complex interplay makes it very difficult to deduce the preceding sequence of operations that lead to a result. Such variability is inherent in a biochemical assay [37]. In this section, we discuss special cases in bioassays that can help the attacker in the reverse-engineering of bio-IP. The following are the cases that aid the attacker.

6.4.1 Fluids with Color

The fluids used for bioassays are usually colorless. However, if the fluid is colored, such as blood serum, then the fluid loading path cannot be obfuscated. Note that distinguishing a color fluid depends on color intensity and imaging precision. If a fluid path meets these criteria, then its path needs to be isolated from the other obfuscated paths. This is done by avoiding ambiguity in the operations (load, mix) involving the colored fluid. We demonstrate this with an example.

Example 26 Consider a 3-inlet multiplexer. Let inlet fluid #2 is colored, and the other two fluids #1, #3 are colorless. Obfuscating the path of fluid #2 is not only futile but counterproductive. When fluid #2 is loaded, any ambiguous actuation will provide clues to the attacker. Therefore, the path (of colored fluid #2) is left unambiguous. Similarly, the subsequent mix operation cannot be obfuscated, i.e., must be left unambiguous.

6.4.2   Fluids with Dispensed Particles

Some inputs have dispensed particles in the fluids. For example, experimental cells or beads are dispensed in a carrier fluid. The loading paths of such fluids have a sieve valve to filter the particles (cells or beads). An additional dummy sieve valve does not obfuscate the loading as the particles could reveal the loading path. Therefore, such a case needs to be treated in the same way as that of the colored fluid path.

6.4.3   Fluids with Sequential Order

An attacker without any knowledge of bio-IP will have to perform trial and error in mapping each fluid input operation. However, an expert attacker with the knowledge of benchtop bio-protocol could predict the order of some of the reagents. For example, MNase reagent is used for DNA digestion in cells, whereas SDS/EDTA lysis buffer is used to arrest the digestion. An attacker with the knowledge of the bio-protocol will be able to predict that the reagent MNase will be used first, followed by SDS/EDTA lysis.

   The obfuscation scheme needs to take into account these special cases to optimize the obfuscation of the bio-IP. The extra valve insertion needs to be done judiciously to maximize the obfuscation effect. Consider a m-inlet multiplexer, let there be x fluids whose order is known, i.e., it is common knowledge that these x fluids are used one after the other. To optimize the obfuscation, we treat the set of x fluids as one fluid. In other words, the net number of inlets is $(m - x + 1)$ for proposes of obfuscation analysis. Using the design rules described for multiplexer, the maximum number of unique ambiguous actuations is $s_{mux} = \binom{m-x+1}{2}$, as shown in Fig. 6.13. Each ambiguous actuation can be mapped to two possibilities. Therefore, the maximum resolution effort is $\mathcal{E}_{mux} = 2^{s_{mux}} = 2^{\binom{m-x+1}{2}}$.

6.5   Experimental Results

In this section, we describe the fabrication of an obfuscated biochip as a proof-of-concept for the proposal. Next, we analyze the application of the obfuscation techniques on a chromatin immunoprecipitation (ChIP) biochip and other real-life biochips.

6.5.1   Fabrication

To demonstrate the practicality of the proposed dummy-valve-based obfuscation, we obfuscated the biochip shown in Fig. 6.1. A 2-layer microfluidic channel master mold was fabricated on a 4 inch Silicon wafer by using conventional softlithography. We created (1) control master mold with multi-height structure and (2) flow master mold. The detailed fabrication process is described in Appendix. The normal valve has a 25 μm thick membrane, whereas the dummy valve was formed with a 50 μm thick membrane. For the applied pressure of *25 psi*, the normal valve closes completely, whereas the dummy valve closes partially, as shown in Fig. 6.4. The resulting CFMB schematic is shown in Fig. 6.14. Note that we have inserted two dummy valves and two extra inputs, i.e., we have applied structural obfuscation. Without the knowledge of the valve type, an attacker cannot infer (1) if the input operation is performed or not and (2) the correct input fluid loaded. This biochip was used to perform successful drug trials, as explained in Example 19. The lab

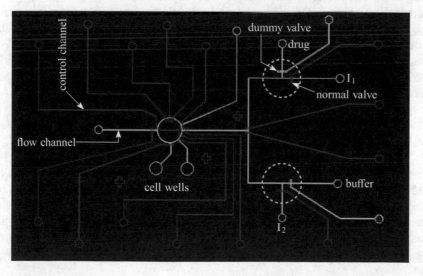

Fig. 6.14 CAD schematic of an obfuscated biochip: The biochip in Fig. 6.1 is obfuscated using dummy inlets and dummy valves

Fig. 6.15 Experimental
setup for drug trial
experiment with an
obfuscated biochip

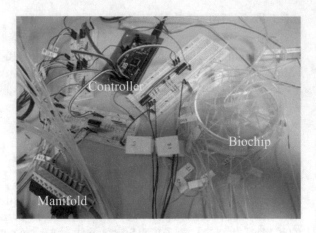

experimental setup is shown in Fig. 6.15. Using the obfuscated biochip, sensitive IP
(drug concentration) can be protected from theft.

6.5.2   Chromatin Immunoprecipitation (ChIP)

The ChIP performs a two-step bioassay: (1) Cell lysis and DNA fragmentation are
performed on the sample cells through a series of mixing operations (Fig. 6.16). This
step uses a 5:1 multiplexer that selects cells and reagents being pushed into the mixer
Ring-1. (2) The resulting fluid is divided equally into four rings (A-D) to perform
ImmunoPrecipitation. These rings are preloaded with antibody functionalized beads
and mixed with cellular material from step 1. Next, the contents of each of the
four rings (A-D) are washed with four different wash buffers (Fig. 6.17a). The
washed beads are then moved to micro-centrifuge tubes for qPCR analysis [135].
We apply reagent load, mix-time, and metering obfuscation to step 1 and structural
obfuscation to step 2 as follows:

6.5.2.1   Load and Mix-Time Obfuscation

One extra valve is added to each inlet of the 5:1 multiplexer, the ring-1's inlet,
and two outlets, i.e., a total of eight extra valves are used, as shown in Fig. 6.16.
This obfuscates the multiplexer selection, the number of mixing operations, and
the mixing time. The maximum number of ambiguous actuations applied to the
multiplexer and mixer are $s_{mux} = \binom{5}{2}$ and $s_{mix} = 4$, respectively. An attacker's
effort in resolving the behavioral obfuscation is $\mathcal{E}_{behav} = \mathcal{E}_{mux} \cdot \mathcal{E}_{mix} = 2^{\binom{5}{2}} \cdot 2^4 = 2^{14}$.

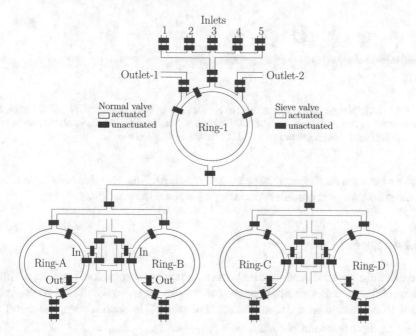

Fig. 6.16 AutoChIP used for gene enrichment. The extra flow channels are shown in blue color

6.5.2.2 Structural Obfuscation

The four ring mixers (A–D) are connected to four fluid inlets that are used to wash the contents of the respective mixer. The inlet channel is replaced by a dummy multiplexer to select between the original wash fluid and a wash fluid corresponding to other mixers, as shown in Fig. 6.16. This results in eight more valves and four more channels. The effort to resolve the structural obfuscation of $n_{dum} = 4$ multiplexers is $\mathcal{E}_{struct} = 2^{n_{dum}} = 2^4$.

6.5.2.3 Metering Obfuscation

We applied metering obfuscation to Ring-1 by transforming the mixer to a 4-segment mixer ($n_p = 4$). Now the reagents can be loaded in variable volumes, i.e., a reagent can be loaded to fill either 0, 1, 2, 3, or 4 segments. Note that the 0-segment loading corresponds to a dummy load operation. This obfuscation requires the addition of 18 valves. The effort to resolve the metering obfuscation is $\mathcal{E}_{meter} = 2^{n_p} = 2^4$. As the cost of this obfuscation is high, we limit it only to Ring-1 and do not apply it to Ring A-D.

The obfuscated sequencing graph is shown in Fig. 6.17b. The effort to resolve the behavioral + structural obfuscation is $\mathcal{E} = \mathcal{E}_{behav} \cdot \mathcal{E}_{struct} \cdot \mathcal{E}_{meter} = 2^{14} \cdot 2^4 \cdot 2^4 = 2^{22}$. Each ChIP trial takes 3.5 h. The time for all trials is over a thousand years. Also,

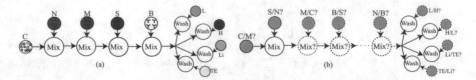

Fig. 6.17 ChIP bioassay: (**a**) original and (**b**) obfuscated. C: Cells under test, N: NP40 buffer, M: Micrococcal nuclease enzyme, S: SDS/EDTA lysis buffer, B: antibody fictionalized beads, L—Low salt buffer, H—High salt buffer, Li—LiCl buffer, and TE—TE buffer

each trial consumes reagents, samples, and biochips. Note that this effort assumes a crude attack; we next consider special cases that aid the attacker.

6.5.2.4   Special Cases

The two fluid inlets of beads and cells have particulate suspension. Loading of these can be obfuscated as a choice between the two. In other words, the 5:1 multiplexer needs to be treated as a 3:1 and a 2:1 multiplexer. In such a case, the effort in resolving the behavioral obfuscation is $\mathcal{E}_{behav} = \mathcal{E}_{mux} \cdot \mathcal{E}_{mix} = 2^{\binom{3}{2}} \cdot 2 \cdot 2^4 = 2^7$. Thereby, the total reverse-engineering effort is given by $\mathcal{E} = \mathcal{E}_{behav} \cdot \mathcal{E}_{struct} \cdot \mathcal{E}_{meter} = 2^7 \cdot 2^4 \cdot 2^4 = 2^{15}$.

Now, let us consider the case where the attacker is aware of the sequence of the reagent. Then, the multiplexer obfuscation is of little use. Then, the effort in resolving the behavioral obfuscation is $\mathcal{E}_{behav} = \mathcal{E}_{mix} = 2^4$. Further, the overall obfuscation is the result of mixer behavior, structural, and metering obfuscations. Thereby, the total reverse-engineering effort is given by $\mathcal{E} = \mathcal{E}_{behav} \cdot \mathcal{E}_{struct} \cdot \mathcal{E}_{meter} = 2^4 \cdot 2^4 \cdot 2^4 = 2^{12}$.

6.5.3   Other Benchmarks

We applied the proposed obfuscation to three more real-life biochips and tabulated the results in Table 6.2. The mRNA iso. and Kinase act. are 4-plex and 2-plex biochips, respectively, where identical assays (attacker trials) are run in parallel. In mRNA iso. ($4 \times 14 = 56$ valves) and Kinase act. ($2 \times 22 = 44$ valves) biochips, $\mathcal{E}$ is smaller due to the replication of a smaller structure. On the other hand, in the larger biochips like ChIP (50 valves) and Nucleic-Acid proc. (54 valves), $\mathcal{E}$ is larger for a comparable design cost in terms of the number of extra valves. The results imply that the dummy-valve-based obfuscation scales well with the complexity of the biochip.

Table 6.2 Crude attack—reverse-engineering of real-life biochips

| Biochip | #valves | # mux | Load | | Mix-time | | | Structural | | | Meter | | | Total effort |
			#extra valves	effort	#mixers	#extra valves	effort	#extra valves	#extra inlets	effort	#extra valves	#extra outlets	effort	
ChIP	50	1 (5:1)	5	2^{10}	4	3	2^4	8	0	2^4	18	2	2^4	2^{22}
Kinase act.	44	2 (3:1)	3	2^3	2	2	2^4	6	3	2^3	36	4	2^4	2^{14}
mRNA iso	56	4 (2:1)	8	2^2	4	8	2^4	8	4	2^4	72	8	2^4	2^{14}
N Acid	54	3 (5:1)	6	2^9	3	6	2^4	6	0	2^3	18	2	2^4	2^{20}

6.5.4 Analysis: Return on Investment

To assess the impact of different types of obfuscation techniques, we define a metric called *return on investment* (RoI). For additional d dummy valves and ports, if the resulting reverse-engineering effort is $\mathcal{E}$, then the return on investment is

$$RoI = \frac{\mathcal{E}}{d} \tag{6.4}$$

An increase in the number of valves leads to an increase in the external solenoid valve and pressure control hardware. It also leads to an increase in the biochip area. However, the cost arising from an increase in the number of valves can be minimized using multiplexer-based addressing of the valves [59].

Type of Obfuscation We tabulate the RoI for different obfuscation techniques applied to the benchmarks in Table 6.3. The results show that with respect to the extra hardware (valves and ports), the best obfuscation type is load, mix-time, structural, and metering, in that order. However, as discussed in the special cases (Sect. 6.4), mix-time and metering obfuscation are more resilient against an attacker who is aware of the sequence of operations. In other words, when the attacker knows the sequence of operations from benchtop bio-protocol, then load and structural obfuscation are ineffective. We show this in Table 6.4, where only bio-protocol parameters mix-time and volume can be obfuscated. Note that valid parameter determination is a critical step during the bio-protocol development process, which requires numerous experimental iterations. Hence, a developer is motivated to protect this critical aspect of the IP in spite of the cost.

Table 6.3 RoI of different obfuscation types

Biochip	ChIP	Kinase act.	mRNA	Nucleic-acid
Behavioral	2048	25.6	4	682.7
Structural	2	0.9	1.3	1.3
Metering	0.8	0.4	0.2	0.8

Table 6.4 Known-sequence attack reverse-engineering effort

| Biochip | Mix-time | | | Meter | | | | Total effort |
	#extra valves	effort	RoI	tiny#extra valves	#extra outlets	effort	RoI	
ChIP	3	2^4	5.3	18	2	2^4	0.8	2^8
Kinase act	2	2^4	8	36	4	22^4	0.4	2^8
mRNA iso.	8	2^4	2	72	8	2^4	0.2	2^8
NA proc.	6	2^4	2.6	18	2	2^4	0.8	2^8

NA: Nucleic-Acid

6.5.5 Comparison Against Other Techniques

The strength of our proposal can be demonstrated in comparison with two IP protection techniques. First, firmware encryption has been used to protect firmware IPs. However, this does not apply to the biochips because the biochip actuations are electrical signals applied to either the valves or to the pneumatic actuators. Even if the actuation sequence is encrypted, it has to be decrypted before it is applied to the biochip control ports. Further, the actuations can be extracted by image and video-based reverse-engineering. The proposed obfuscation complements encryption to thwart reverse-engineering of the electrical signals. Second, logic locking is used to prevent IP piracy in VLSI designs. The number of trials needed to de-obfuscate a logic-locked VLSI design is of the order of 2^{128} [108]. These trials can be done on high-speed computers. On the other hand, the bioassay trials take several hours to complete. Also, unlike VLSI, the bioassay recovery trials require perishable reagents and biochips. The cost and time spent on these trials go against an attacker's economic objective of stealing a bioassay IP.

6.6 Conclusion

Microfluidic platforms have immense potential in paving the way for rapid and low-cost biochemical analysis. However, the cyberphysical system that enables biochip automation is susceptible to IP theft. This is a major hurdle in the large-scale adaptation of microfluidic technologies in industries that are prone to stealing sensitive research data. Our work addresses this pressing problem with a practical obfuscation methodology that can be easily integrated with the current biochip design flow. We developed dummy valve based obfuscation design rules and showcased their application to real-life biochips. The results show that the de-obfuscation effort is daunting enough to act as a deterrent to an attacker.

The strength of the obfuscation can be demonstrated in comparison with two IP protection techniques. First, firmware encryption has been used to protect firmware IPs. However, this does not apply to the biochips because the biochip actuations are electrical signals applied to either the valves or to the pneumatic actuators. Even if the actuation sequence is encrypted, it has to be decrypted before it is applied to the biochip control ports. Further, the actuations can be extracted by image and video-based RE. Proposed obfuscation complements encryption to thwart RE of the electrical signals. Second, logic locking is used to prevent IP piracy in VLSI designs. The number of trials needed to de-obfuscate a logic-locked VLSI design is of the order of 2^{128} [108]. These trials can be done on high-speed computers. On the other hand, the bioassay trials take several hours to complete. Also, unlike VLSI, the bioassay recovery trials require perishable reagents and biochips. The cost and time spent on these trials go against an attacker's economic objective of stealing a bioassay IP.

Future Outlook

Biochip CPS is gaining prominence as it offers several advantages over traditional biochemical labs. Biochips use microfluidic technologies that enable manipulation of the nano-/pico-liter volume of fluids on a miniaturized platform. Biochips use minimal sample volumes, are quicker, and can be automated. These advantages enable new frontiers. Incidentally, biochips are targeted for safety-critical applications. However, multiple stages of the supply chain are susceptible to various attacks. Gaining the trust of investors, regulators, and users is critical to the large-scale deployment of biochips.

This book focuses on security and IP protection in biochip cyberphysical systems (CPS). It outlines security evaluation, trust enhancement, and IP protection in biochip CPS. The use of emerging devices in CPS demands that we relook at the traditional safety and reliability measures in the CPS. These emerging applications have different computational and energy resources, and therefore, the safety measures of general-purpose computing cannot be applied as-is. Biochip security overlaps the fields of hardware and software controls, biology, security, and IP protection. This book attempts to bridge this gap by evaluating threats to CPS in emerging contexts and devising countermeasures that leverage the available resources.

Research work has so far focused on biochip integrity and IP protection in digital and flow-based biochips. We have successfully adapted several well-established security paradigms to biochips, such as obfuscation, watermarking, exact analysis, ML-based attack detection. This involved a careful understanding of the underlying technology and system constraints. The results have been very encouraging. The future security work needs to further this progress by focusing on issues and solutions that are unique or specific only to microfluidic technology. Towards this end, the following can be possible leads.

Side-Channel Assessment

The vulnerability of a biochip CPS to novel physical and side-channel based attacks needs to be probed. In cryptographic systems, E&M and power side-channels have been shown to leak information. Biochip may have new unanticipated side-channels such as audio or electromagnetic. This can be used to leak information on the biochemical processes on the biochip, such as diagnostic results. Investigation is needed on applications of microfluidic platforms in research settings to uncover the potential misinformation that can be generated from an attack.

On the other hand, side-channel information has been used in integrated circuit design to detect Trojans and establish the correctness of the design. The biochip side-channels can be used for signature-based online monitoring. For example, the solenoid valve interface to flow-based biochips generates audio side-channel during the opening/closing of valves. Each solenoid valve can have different audio associated with it. This can be used as a signature-based verification of design implementation.

Conditional Bioassay Security

A bioassay can include conditional loops such as if-then-else statements or loops with a non-constant number of iterations. In such a case, the actual execution path is resolved at run-time based on sensor readings. So far, the security analysis of biochip systems assumes the static baseline implementation of the bioassay. However, many real-world biochemical protocols require conditional execution, where certain output droplets determine the overall execution flow of the sequencing graph [74]. There is a need to explore the vulnerabilities in such control path-dominated designs. For example, the checkpointing mechanism assumes a fixed static execution of bioassay, whereas, in fact, it has a dynamic execution in the presence of hardware errors. We need to develop algorithms that respond to conditional execution and develop a dynamic checkpointing scheme. Such a scheme needs to ensure (1) the dynamic execution is one of the valid execution and (2) it is consistent with the hardware errors reported.

Similarly, the checkpoint verification method does not consider dynamic execution. A possible solution can be to save a set of predetermined checkpoints depending on the execution path and use the SAT solver offline to verify the execution. Our current solution can be used to derive such predetermined checkpoints for each possible execution path. During run-time, the checkpoints corresponding to the chosen path are captured and saved for verification.

Assessing Commercial Devices

Having demonstrated the attacks in simulation, the next step is to demonstrate the tampering attacks on the commercial biochip products. The attacks can be simple household attack procedures. The effectiveness of attack is captured by its impact on the end result and the traces left by the attack. Moreover, a systematic procedure needs to be laid out for reverse-engineering the commercial product. This attack methodology, then, can be adapted to commercial DMFBs and CFMBs.

User-Verifiable Security

To increase the trust in biochip systems, a layer of security needs to be provided that can be applied to diverse usage scenarios such as in a biomedical lab and in a remote online/offline diagnostic device. This can be done by adding visual droplet checks that verify the intermediate droplet status. In other words, the end-user can inspect the color of the intermediate droplets and verify if the expected operations have been executed. This way, we achieve a low-cost, versatile verification scheme.

Design for Security

As future work, we plan to devise algorithms for the automatic synthesis of a bioprotocol, which is provably secure, given a checkpoint defense mechanism. Synthesis algorithms currently do not take into consideration the security needs of the application [74]. Previous work on checkpoint-based defenses has shown that the biochip systems are resource-constrained to implement a provably secure defense [121]. A top-down approach of considering security requirements before synthesis will help in developing practical fool-proof security. This can be achieved in two ways:

1. Given bioassay's throughput constraint in terms of the number of cycles, the synthesis tool realizes an execution and gives the required computational support for security in terms of the max number of checkpoints per cycle.
2. Given biochip's computational constraint in terms of the max number of checkpoints per cycle, the synthesis tool realizes a secure execution and gives the throughput cost in terms of the number of cycles for bioassay completion.

There is a need to devise a testing strategy to determine the valid pressure range that allows a reliable valve operation. We can use a valve simulation model to predict the ranges and then confirm the same during the test [79]. As rigorous testing might spoil the biochip, a scheme is required for mapping each valve to an external solenoid manifold, such that the pressure fluctuations are kept within the expected range. This should help in avoiding Trojan triggers.

Bibliography

1. 5 cases of AIDS-study sabotage reported, 1986. https://www.chicagotribune.com/news/ct-xpm-1986-09-18-8603090884-story.html
2. It's insanely easy to hack hospital equipment, 2014. https://www.wired.com/2014/04/hospital-equipment-vulnerable/
3. Keccak Hash Function, NIST, 2014. http://csrc.nist.gov/groups/ST/hash/sha-3
4. 2 GSK scientists indicted in secrets case involving China, 2016. https://www.justice.gov/usao-edpa/pr/scientists-indicted-allegedly-stealing-biopharmaceutical-trade-secrets
5. Drug development and intellectual property theft, 2016. https://digitalguardian.com/blog/drug-development-and-intellectual-property-theft
6. Fda advisors back approval of baebies' seeker analyzer for newborns, 2016. http://baebies.com/fda-advisors-back-approval-baebies-seeker-analyzer-newborns
7. A high throughput screening system to identify actionable treatments for cancer patients, 2016. https://biosero.com/wp-content/uploads/2016/10/a_high_throughput_screening_system_to_identify_actionable_treatments_etc_npm.pdf
8. NIST, 2016. https://www.keylength.com/en/4/
9. Wheeler group at kakuma refugee camp in kenya, 2016. http://www.chem.utoronto.ca/edistillations/fall2016/kakuma.html
10. Pharmaceutical giant rocked by ransomware attack, 2017. https://www.washingtonpost.com/news/the-switch/wp/2017/06/27/pharmaceutical-giant-rocked-by-ransomware-attack/
11. Automating lab protocols, 2018. https://www.illumina.com/science/technology/digital-microfluidics.html
12. Ella platform, ProteinSimple, 2018. https://www.proteinsimple.com/ella.html
13. Fluidigm systems, 2018. https://www.fluidigm.com/systems
14. IBM ILOG CPLEX Optimizer, 2018. http://www.ibm.com/software/integration/optimization/cplex/
15. Illumina-press release, 2018. https://www.illumina.com/company/news-center/press-releases/press-release-details.html?newsid=1840193
16. JOVE - The ELISA Method, 2018. https://www.jove.com/science-education/5061/the-elisa-method
17. Laboratory monitoring, 2018. http://tetrascience.com/case-studies/laboratory-monitoring-notable-labs
18. Sunrise, TECAN, 2018. https://lifesciences.tecan.com/products/microplate_readers
19. Understanding variability in DNA amplification reactions, 2018. https://www.neb.com/tools-and-resources/feature-articles/understanding-variability-in-dna-amplification-reactions

20. When medical devices get hacked, hospitals often don't know it., 2018. https://www.health careitnews.com/news/when-medical-devices-get-hacked-hospitals-often-dont-know-it

21. 10x genomics funding, 2019. https://www.10xgenomics.com/news/10x-genomics-lands-new-financing/

22. Fluidigm Q1 revenue 2019, 2019. http://investors.fluidigm.com/news-releases/news-release-details/fluidigm-announces-first-quarter-2019-financial-results

23. Jealousy led montana chemist to taint colleague's water tests, 2019. https://www.nytimes.com/2019/08/08/us/montana-chemist-water.html

24. SEngine - Precision medicine, 2019. https://senginemedicine.com/

25. Shipment of 3,000,000th test, 2019. https://baebies.com/celebrating-shipment-of-3000000th-test/

26. Theranos effect, 2019. https://www.bloomberg.com/news/articles/2019-03-23/elizabeth-holmes-theranos-still-haunt-blood-testing-startups

27. Theranos voids two years of edison blood-test results, 2019. https://www.wsj.com/articles/theranos-voids-two-years-of-edison-blood-test-results-1463616976

28. Zion market research, 2019. https://www.globenewswire.com/news-release/2019/04/17/1805498/0/en/Global-Microfluidics-Market-Will-Surpass-USD-12-380-Million-By-2025-Zion-Market-Research.html

29. G.-E. A, S. Brahim, and G. Slaughter. Design of a subcutaneous implantable biochip for monitoring of glucose and lactate. *IEEE Sensors J.*, 5(3):345–355, 2005.

30. J. K. Actor. *Elsevier's Integrated Review Immunology and Microbiology*. Elsevier Inc., 2012.

31. S. S. Ali, M. Ibrahim, J. Rajendran, O. Sinanoglu, and K. Chakrabarty. Supply-chain security of digital microfluidic biochips. *IEEE Computer*, 49(8):36–43, 2016.

32. S. S. Ali, M. Ibrahim, O. Sinanoglu, K. Chakrabarty, and R. Karri. Security implications of cyberphysical digital microfluidic biochips. In *International Conference Computer Design*, pages 483–486, 2015.

33. S. S. Ali, M. Ibrahim, O. Sinanoglu, K. Chakrabarty, and R. Karri. Microfluidic encryption of on-chip biochemical assays. In *IEEE Biomed. Circuits Systems Conference*, pages 152–155, 2016.

34. Q. A. Arain, D. Zhongliang, I. Memon, S. Arain, F. K. Shaikh, A. Zubedi, M. A. Unar, A. Ashraf, and R. Shaikh. Privacy preserving dynamic pseudonym-based multiple mix-zones authentication protocol over road networks. *Wireless Personal Commun.*, 95(2):505–521, July 2017.

35. S. Bhattacharjee, A. Banerjee, K. Chakrabarty, and B. B. Bhattacharya. Correctness checking of bio-chemical protocol realizations on a digital microfluidic biochip. In *Proceedings of VLSID*, pages 504–509, 2014.

36. S. Bhattacharjee, S. Chatterjee, A. Banerjee, T. Ho, K. Chakrabarty, and B. B. Bhattacharya. Adaptation of biochemical protocols to handle technology-change for digital microfluidics. *IEEE Transactions on on CAD of Integrated Circuits and Systems*, 36(3):370–383, 2017.

37. S. Bhattacharjee, Y.-L. Chen, J.-D. Huang, and B. B. Bhattacharya. Concentration-resilient mixture preparation with digital microfluidic lab-on-chip. *ACM Transactions on Embed. Computer Systems*, 17(2):49:1–49:12, Jan. 2018.

38. S. Bhattacharjee, J. Tang, M. Ibrahim, K. Chakrabarty, and R. Karri. Locking of biochemical assays for digital microfluidic biochips. In *European Test Symposium*, pages 1–6, 2018.

39. B. B. Bhattacharya, S. Roy, and S. Bhattacharjee. Algorithmic challenges in digital microfluidic biochips: Protocols, design, and test. In *Proceedings of International Conference Applied Algorithms*, pages 1–16, 2014.

40. S. Chakraborty, C. Das, and S. Chakraborty. Securing module-less synthesis on cyberphysical digital microfluidic biochips from malicious intrusions. In *Proceedings of VLSID*, pages 467–468, 2018.

41. S. Chakraborty, C. Das, and S. Chakraborty. Securing module-less synthesis on cyberphysical digital microfluidic biochips from malicious intrusions. In *Intl Conference VLSI Design VLSID*, pages 467–468, Jan 2018.

42. H. Chen, S. Potluri, and F. Koushanfar. Biochipwork: Reverse engineering of microfluidic biochips. In *Proceedings of ICCD*, pages 9–16, 2017.

43. H. Chen, S. Potluri, and F. Koushanfar. Biochipwork: Reverse engineering of microfluidic biochips. In *International Conference on Computer Design*, pages 9–16, 2017.

44. X. Chen, G. Qu, and A. Cui. Practical ip watermarking and fingerprinting methods for asic designs. In *Proc. Int. Symposium on Circuits Systems*, pages 1–4. IEEE, 2017.

45. C. D. Chin, V. Linder, and S. K. Sia. Commercialization of microfluidic point-of-care diagnostic devices. *Lab Chip*, 12:2118–2134, 2012.

46. K. Choi, A. H. C. Ng, R. Fobel, D. A. Chang-Yen, L. E. Yarnell, E. L. Pearson, C. M. Oleksak, A. T. Fischer, R. P. Luoma, J. M. Robinson, J. Audet, and A. R. Wheeler. Automated digital microfluidic platform for magnetic-particle-based immunoassays with optimization by design of experiments. *Anal. Chem.*, 85(20):9638–9646, 2013.

47. S. Chung, J. Park, C. Chung, D. C. Han, and J. K. Chang. Multi-height micro structures in poly (dimethyl siloxane) lab-on-a-chip. *MicroSystems Technology*, 10(2):81–88, Jan. 2004.

48. C. Collberg and C. Thomborson. Software watermarking: Models and dynamic embeddings. In *Proc. ACM SIGPLAN-SIGACT Symposium Principles of Programming Languages*, POPL '99, pages 311–324. ACM, 1999.

49. I. J. Cox, J. Kilian, F. T. Leighton, and T. Shamoon. Secure spread spectrum watermarking for multimedia. *IEEE Transactions of Image Process*, 6(12):1673–1687, 1997.

50. I. J. Cox, M. L. Miller, and J. A. Bloom. *Digital Watermarking*. Morgan Kaufmann, 2001.

51. K. L. Cox, V. Devanarayan, A. Kriauciunas, J. Manetta, C. Montrose, and S. Sittampalam. Immunoassay methods. *Assay Guidance Manual [Internet]*, 2014.

52. A. Cui, C.-H. Chang, S. Tahar, and A. T. Abdel-Hamid. A robust fsm watermarking scheme for ip protection of sequential circuit design. *IEEE Transactions on Computer-Aided Design Integrated Circuits and Systems*, 30(5):678–690, 2011.

53. L. M. de Moura and N. Bjørner. Z3: An efficient SMT solver. In *Proceedings of TACAS*, pages 337–340, [Z3 is available at https://github.com/Z3Prover/z3], 2008.

54. N. Dey, A. S. Ashour, F. Shi, S. J. Fong, and J. M. R. S. Tavares. Medical cyber-physical systems: A survey. *Journal of Medical Systems*, 42(4):74, 2018.

55. C. Dong, L. Liu, H. Liu, W. Guo, X. Huang, S. Lian, X. Liu, and T. Ho. A survey of dmfbs security. State-of-the-art attack and defense. In *2020 21st International Symposium on Quality Electronic Design (ISQED)*, pages 14–20, 2020.

56. R. B. Fair. Digital microfluidics: is a true lab-on-a-chip possible? *Microfluid Nanofluid*, 3(3):245–281, 2007.

57. R. B. Fair, A. Khlystov, T. Tailor, V. Ivanov, R. Evans, V. Srinivasan, V. Pamula, M. Pollack, P. Griffin, and J. Zhou. Chemical and biological applications of digital-microfluidic devices. *IEEE Design & Test of Computers*, 24(1):10–24, 2007.

58. O. S. Faragallah. Secure audio cryptosystem using hashed image lsb watermarking and encryption. *Wirel. Pers. Commun.*, 98(2):2009–2023, Jan. 2018.

59. L. M. Fidalgo and S. J. Maerkl. A software-programmable microfluidic device for automated biology. *Lab Chip*, 11:1612–1619, 2011.

60. R. Fobel, C. Fobel, and A. R. Wheeler. DropBot: An open-source digital microfluidic control system with precise control of electrostatic driving force and instantaneous drop velocity measurement. *Applied Physics Letters*, 102(19):193513, 2013.

61. R. W. Forsman. Why is the laboratory an afterthought for managed care organizations? *Clinical Chemistry*, 42(5):813–816, 1996.

62. S. A. M. Gilani, I. Kostopoulos, and A. N. Skodras. Color image-adaptive watermarking. In *Int. Conference Digital Signal Process. Proc.*, volume 2, pages 721–724, 2002.

63. C. C. Glick, M. T. Srimongkol, A. J. Schwartz, W. S. Zhuang, J. C. Lin, R. H. Warren, D. R. Tekell, P. A. Satamalee, and L. Lin. Rapid assembly of multilayer microfluidic structures via 3D-printed transfer molding and bonding. *Microsystems & Nanoengineering*, 2:16063, 2016.

64. A. Grimmer, Q. Wang, H. Yao, T. Y. Ho, and R. Wille. Close-to-optimal placement and routing for continuous-flow microfluidic biochips. In *Asia and South Pacific Design Automation Conf*, pages 530–535, 2017.

65. D. Grissom, C. Curtis, S. Windh, S. Phung, N. Kumar, Z. Zimmerman, K. O'Neal, J. McDaniel, N. Liao, and P. Brisk. An open-source compiler and pcb synthesis tool for digital microfluidic biochips. *Integration: The VLSI Journal*, 51:169–193, 2015.

66. D. Grissom, K. O'Neal, B. Preciado, H. Patel, R. Doherty, N. Liao, and P. Brisk. A digital microfluidic biochip synthesis framework. In *Proceedings of International Conference VLSI*, pages 177–182, 2012.

67. W. H. Grover, R. H. C. Ivester, E. C. Jensen, and R. A. Mathies. Development and multiplexed control of latching pneumatic valves using microfluidic logical structures. *Lab Chip*, 6:623–631, 2006.

68. J. Guo and M. Potkonjak. Watermarking deep neural networks for embedded systems. In *2018 IEEE/ACM Int. Conference on Computer-Aided Design*, pages 1–8, Nov 2018.

69. T. Hastie, R. Tibshirani, and J. Friedman. *The Elements of Statistical Learning*. Springer New York Inc., New York, NY, USA, 2001.

70. T.-Y. Ho, K. Chakrabarty, and P. Pop. Digital microfluidic biochips: recent research and emerging challenges. In *IEEE/ACM International Conference Hardware/Software Codesign Syst Synthesis.*, pages 335–344, 2011.

71. K. Hu, M. Ibrahim, L. Chen, Z. Li, K. Chakrabarty, and R. Fair. Experimental demonstration of error recovery in an integrated cyberphysical digital-microfluidic platform. In *Proceedings of IEEE BioCAS.*, pages 1–4, 2015.

72. K. Hu, F. Yu, T. Ho, and K. Chakrabarty. Testing of flow-based microfluidic biochips: Fault modeling, test generation, and experimental demonstration. *IEEE Transactions on Computer-Aided Design of Integrated Circuits and Systems*, 33(10):1463–1475, 2014.

73. J.-D. Huang, C. Liu, and T. Chiang. Reactant minimization during sample preparation on digital microfluidic biochips using skewed mixing trees. In *Proceedings of ICCAD*, pages 377–383, 2012.

74. M. Ibrahim, K. Chakrabarty, and K. Scott. Synthesis of cyberphysical digital-microfluidic biochips for real-time quantitative analysis. *IEEE Transactions on Computer-Aided Design Integrated Circuits Systems*, 36(5):733–746, 2017.

75. N. Japkowicz and M. Shah. *Evaluating Learning Algorithms: A Classification Perspective*. Cambridge University Press, New York, NY, USA, 2011.

76. O. Keszocze, Z. Li, A. Grimmer, R. Wille, K. Chakrabarty, and R. Drechsler. Exact routing for micro-electrode-dot-array digital microfluidic biochips. In *Proceedings of ASPDAC*, pages 708–713, 2017.

77. N. Khalid, I. Kobayashi, and M. Nakajima. Recent lab-on-chip developments for novel drug discovery. *Wiley Interdiscip. Rev. Systems Biol. Med.*, 9(4):e1381, 2017.

78. R. Langner. Stuxnet: dissecting a cyberwarfare weapon. *IEEE Security Privacy*, 9(3):49–51, 2011.

79. A. Lau, H. Yip, K. Ng, X. Cui, and R. Lam. Dynamics of microvalve operations in integrated microfluidics. *Micromachines*, 5(1):50–65, 2014.

80. R. M. Lequin. Enzyme immunoassay (eia)/enzyme-linked immunosorbent assay (elisa). *Clinical Chemistry*, 51(12):2415–2418, 2005.

81. Z. Li, T.-Y. Ho, K. Lai, K. Chakrabarty, P. Yu, and C. Lee. High-level synthesis for micro-electrode-dot-array digital microfluidic biochips. In *Proceedings of DAC*, pages 1–6, 2016.

82. Z. Li, K. Lai, P.-H. Yu, K. Chakrabarty, T.-Y. Ho, and C. Lee. Droplet size-aware high-level synthesis for micro-electrode-dot-array digital microfluidic biochips. *IEEE Transactions on Biomed. Circuits Systems*, 11(3):612–626, 2017.

83. Z. Li, K. Y. Lai, P.-H. Yu, K. Chakrabarty, M. Pajic, T.-Y. Ho, and C. Lee. Error recovery in a micro-electrode-dot-array digital microfluidic biochip. In *Proceedings of ICCAD*, pages 1–8, 2016.

84. Z. Li, K. Y.-T. Lai, J. McCrone, P.-H. Yu, K. Chakrabarty, M. Pajic, T.-Y. Ho, and C.-Y. Lee. Efficient and adaptive error recovery in a micro-electrode-dot-array digital microfluidic biochip. *IEEE Transactions on Computer-Aided Design Integrated Circuits Systems*, 99, 2017.

85. T.-C. Liang, M. Shayan, K. Chakrabarty, and R. Karri. Execution of provably secure assays on meda biochips to thwart attacks. In *Proceedings of the 24th Asia and South Pacific Design Automation Conference*, pages 51–57, 2019.

86. T.-C. Liang, M. Shayan, K. Chakrabarty, and R. Karri. Execution of provably secure assays on meda biochips to thwart attacks. In *Proceedings of ASPDAC*, pages 51–57, 2019.

87. C. Lin, J.-D. Huang, H. Yao, and T.-Y. Ho. A comprehensive security system for digital microfluidic biochips. In *Proc ITC Asia*, pages 151–156, 2018.

88. C. Liu, B. Li, B. B. Bhattacharya, K. Chakrabarty, T.-Y. Ho, and U. Schlichtmann. Testing microfluidic fully programmable valve arrays (FPVAs). In *Design, Auto. Test in Europe*, pages 91–96, 2017.

89. Y. Luo, K. Chakrabarty, and T.-Y. Ho. Error recovery in cyberphysical digital microfluidic biochips. *IEEE Transactions on on CAD of Integrated Circuits and Systems*, 32(1):59–72, 2013.

90. Y.-H. V. Ma, K. Middleton, L. You, and Y. Sun. A review of microfluidic approaches for investigating cancer extravasation during metastasis. *Microsystems & Nanoengineering*, 4(17104):1–13, 2018.

91. J. Melin and S. R. Quake. Microfluidic large-scale integration: The evolution of design rules for biological automation. *Annual Review of Biophysics and Biomolecular Structure*, 36(1):213–231, 2007.

92. S. Mohammed, S. Bhattacharjee, T.-C. Liang, J. Tang, K. Chakrabarty, and R. Karri. Shadow attacks on MEDA biochips. In *International Conference Computer Aided Design*, pages 73:1–8, 2018.

93. S. Mohammed, S. Bhattacharjee, Y. A. Song, K. Chakrabarty, and R. Karri. Desieve the attacker: Thwarting ip theft in sieve-valve-based biochips. In *Design, Auto. Test in Europe*, pages 210–215, 2019.

94. S. Mohammed, S. Bhattacharjee, Y. A. Song, K. Chakrabarty, and R. Karri. Security assessment of microfluidic fully-programmable-valve-array biochips. In *International Conference of VLSI Design*, pages 197–202, 2019.

95. S. Mohammed, S. Bhattacharjee, Y. A. Song, K. Chakrabarty, and R. Karri. Can multi-layer microfluidics aid bio-intellectual property protection? In *International Symposium on On-line Test and Robust Systems Design*, 2019 (to appear).

96. S. Mohammed, S. Bhattacharjee, Y. A. Song, K. Chakrabarty, and R. Karri. Desieve the attacker: Thwarting ip theft in sieve-valve-based biochips. In *Design, Autom. Test in Europe*, 2019 (to appear).

97. S. Mohammed, J. Tang, K. Chakrabarty, and R. Karri. Security assessment of microelectrode-dot-array biochips. *IEEE Transactions on Computer-Aided Design Integrated Circuits Systems*, 2018 (to appear).

98. K. P. Murphy. *Machine Learning: A Probabilistic Perspective*. The MIT Press, 2012.

99. A. H. C. Ng, K. Choi, R. P. Luoma, J. M. Robinson, and A. R. Wheeler. Digital microfluidic magnetic separation for particle-based immunoassays. *Anal. Chem.*, 84(20):8805–8812, 2012.

100. A. H. C. Ng, R. Fobel, C. Fobel, J. Lamanna, D. G. Rackus, A. Summers, C. Dixon, M. D. M. Dryden, C. Lam, M. Ho, N. S. Mufti, V. Lee, M. A. M. Asri, E. A. Sykes, M. D. Chamberlain, R. Joseph, M. Ope, H. M. Scobie, A. Knipes, P. A. Rota, N. Marano, P. M. Chege, M. Njuguna, R. Nzunza, N. Kisangau, J. Kiogora, M. Karuingi, J. W. Burton, P. Borus, E. Lam, and A. R. Wheeler. A digital microfluidic system for serological immunoassays in remote settings. *Science Translational Medicine*, 10(438), 2018.

101. A. H. C. Ng, U. Uddayasankar, and A. R. Wheeler. Immunoassays in microfluidic systems. *Anal. Bioanal.Chem.*, 397(3):991–1007, Jun 2010.

102. A. H. C. Ng, U. Uddayasankar, and A. R. Wheeler. Immunoassays in microfluidic systems. *Anal. Bioanal.Chem.*, 397(3):991–1007, Jun 2010.

103. U. D. of Homeland Security and CERT. Russian government cyber activity targeting energy and other critical infrastructure sectors, 2018.

104. F. Pedregosa, G. Varoquaux, A. Gramfort, V. Michel, B. Thirion, O. Grisel, M. Blondel, P. Prettenhofer, R. Weiss, V. Dubourg, J. Vanderplas, A. Passos, D. Cournapeau, M. Brucher, M. Perrot, and E. Duchesnay. Scikit-learn: Machine learning in Python. *Journal of Machine Learning Research*, 12:2825–2830, 2011.

105. P. Pop, I. E. Araci, and K. Chakrabarty. Continuous-flow biochips: Technology, physical-design methods, and testing. *IEEE Design & Test*, 32(6):8–19, 2015.

106. M. Potkonjak, G. Qu, F. Koushanfar, and C.-H. Chang. 20 years of research on intellectual property protection. In *Proceedings of IEEE International Symposium Circuits Systems*, pages 1–4. IEEE, 2017.

107. S. R. Quake, J. S. Marcus, and C. L. Hansen. Microfluidic sieve valves, Jan 2015. US Patent 8932461B2.

108. J. Rajendran, H. Zhang, C. Zhang, G. S. Rose, Y. Pino, O. Sinanoglu, and R. Karri. Fault analysis-based logic encryption. *IEEE Transactions of on Computers*, 64:410–424, 2015.

109. J. Riordon, D. Sovilj, S. Sanner, D. Sinton, and E. W. Young. Deep learning with microfluidics for biotechnology. *Trends in Biotechnology*, 37(3):310–324, 2019.

110. U. S. Service, CERT, C. Magazine, and Deloitte. 2011 cybersecurity watch survey: How bad is the insider threat?, 2011.

111. M. Shayan, S. Bhattacharjee, T.-C. Liang, J. Tang, K. Chakrabarty, and R. Karri. Shadow attacks on meda biochips. In *Proceedings International Conference on Computer-Aided Design*, pages 73:1–73:8, 2018.

112. M. Shayan, S. Bhattacharjee, A. Orozaliev, Y.-A. Song, K. Chakrabarty, and R. Karri. Thwarting bio-ip theft through dummy-valve-based obfuscation. *IEEE Transactions on Information Forensics and Security*, 16:2076–2089, 2021.

113. M. Shayan, S. Bhattacharjee, Y. Song, K. Chakrabarty, and R. Karri. Toward secure microfluidic fully programmable valve array biochips. *IEEE Transactions on Very Large Scale Integration (VLSI) Systems*, pages 1–12, to appear 2019.

114. M. Shayan, S. Bhattacharjee, J. Tang, K. Chakrabarty, and R. Karri. Bio-protocol watermarking on digital microfluidic biochips. *IEEE Transactions on on Information Forensics and Security*, 2019.

115. M. Shayan, S. Bhattacharjee, R. Wille, K. Chakrabarty, and R. Karri. How secure are checkpoint-based defenses in digital microfluidic biochips? *IEEE Transactions on Computer-Aided Design of Integrated Circuits and Systems*, 40(1):143–156, 2021.

116. M. Shayan, T.-C. Liang, S. Bhattacharjee, K. Chakrabarty, and R. Karri. Toward secure checkpointing for micro-electrode-dot-array biochips. *IEEE Transactions on Computer-Aided Design of Integrated Circuits and Systems*, 39(12):4908–4920, 2020.

117. M. Shayan, J. Tang, K. Chakrabarty, and R. Karri. Security assessment of micro-electrode-dot-array biochips. *IEEE Transactions on on CAD of Integrated Circuits and Systems*, 2018 (early access).

118. R. Sista, Z. Hua, P. Thwar, A. Sudarsan, V. Srinivasan, A. Eckhardt, M. Pollack, and V. Pamula. Development of a digital microfluidic platform for point of care testing. *Lab Chip*, 8(12):2091–2104, 2008.

119. J. Tang, M. Ibrahim, K. Chakrabarty, and R. Karri. Security implications of cyberphysical flow-based microfluidic biochips. In *IEEE Asian Test Symposium*, pages 115–120, 2017.

120. J. Tang, M. Ibrahim, K. Chakrabarty, and R. Karri. Security trade-offs in microfluidic routing fabrics. In *Proceedings of ICCD*, pages 25–32, 2017.

121. J. Tang, M. Ibrahim, K. Chakrabarty, and R. Karri. Secure randomized checkpointing for digital microfluidic biochips. *IEEE Transactions on Computer-Aided Design Integrated Circuits Systems*, 37(6):1119–11132, 2018.

122. J. Tang, M. Ibrahim, K. Chakrabarty, and R. Karri. Tamper-resistant pin-constrained digital microfluidic biochips. In *Proceedings of DAC*, pages 1–6, 2018.

123. J. Tang, M. Ibrahim, K. Chakrabarty, and R. Karri. Towards secure and trustworthy cyberphysical microfluidic biochips. *IEEE Transactions on on CAD of Integrated Circuits and Systems*, 2018 (early access).

124. J. Tang, R. Karri, M. Ibrahim, and K. Chakrabarty. Securing digital microfluidic biochips by randomizing checkpoints. In *International Test Conference*, pages 1–8, 2016.

125. J. Tang, R. Karri, M. Ibrahim, and K. Chakrabarty. Tamper-resistant pin-constrained digital microfluidic biochips. In *Design Automation Conference*, pages 67:1–67:6, 2018.

126. W. Thies, J. P. Urbanski, T. Thorsen, and S. P. Amarasinghe. Abstraction layers for scalable microfluidic biocomputing. *Natural Computing*, 7(2):255–275, 2008.

127. T.-M. Tseng, B. Li, T.-Y. Ho, and U. Schlichtmann. Reliability-aware synthesis for flow-based microfluidic biochips by dynamic-device mapping. In *Design Automation Conference*, pages 1–6, 2015.

128. M. Turetta, F. D. Ben, G. Brisotto, E. Biscontin, M. Bulfoni, D. Cesselli, A. Colombatti, G. Scoles, G. Gigli, and L. L. Del Mercato. Emerging technologies for cancer research: Towards personalized medicine with microfluidic platforms and 3d tumor models. *Current Medicinal Chemistry*, 25:4616–4637, 2018.

129. L. F. Turner. Digital data security system, 1989.

130. J. P. Urbanski, W. Thies, C. Rhodes, S. Amarasinghe, and T. Thorsen. Digital microfluidics using soft lithography. *Lab Chip*, 6:96–104, 2006.

131. U.S.Government. Increase in insider threat cases highlight significant risks to business networks and proprietary information, 2014.

132. N. Vergauwe, D. Witters, F. Ceyssens, S. Vermeir, B. Verbruggen, R. Puers, and J. Lammertyn. A versatile electrowetting-based digital microfluidic platform for quantitative homogeneous and heterogeneous bio-assays. *J. Micromech. Microeng.*, 21(5):054026, 2011.

133. G. Wang, D. Teng, and S.-K. Fan. Digital microfluidic operations on micro-electrode array architecture. In *Proceedings of International Conference on Nano/Micro Engineered and Molecular Systems*, 2011.

134. G. M. Whitesides. The origins and the future of microfluidics. *Nature*, 442(7101):368–373, 2006.

135. A. R. Wu, J. B. Hiatt, R. Lu, J. L. Attema, N. A. Lobo, I. L. Weissman, M. F. Clarke, and S. R. Quake. Automated microfluidic chromatin immunoprecipitation from 2,000 cells. *Lab Chip*, 9:1365–1370, 2009.

136. D. Yeung, S. Ciotti, S. Purushothama, E. Gharakhani, G. Kuesters, B. Schlain, C. Shen, D. Donaldson, and A. Mikulskis. Evaluation of highly sensitive immunoassay technologies for quantitative measurements of sub-pg/ml levels of cytokines in human serum. *Journal of Immunological Methods*, 437:53–63, 2016.

137. K. Yi-Tse Lai, Y.-T. Yang, and C. Y. Lee. An intelligent digital microfluidic processor for biomedical detection. *J. Signal Process. Systems*, 78(1):85–93, 2015.

138. J. Zambreno, A. Choudhary, R. Simha, B. Narahari, N. Memon, and N. Memon. Safe-ops: An approach to embedded software security. *ACM Transactions on Embed. Computer Systems*, 4(1):189–210, Feb. 2005.

139. Y. S. Zhang. A medical mini-me: one day your doctor could prescribe drugs based on now a biochip version of you reacts. *IEEE Spectrum*, 56(4):44–49, April 2019.

140. Y. Zhao, T. Xu, and K. Chakrabarty. Integrated control-path design and error recovery in the synthesis of digital microfluidic lab-on-chip. *ACM J. Emerging Technology Computer Systems*, 6(3):1–28, 2010.

141. Z. Zhong, Z. Li, and K. Chakrabarty. Adaptive error recovery in meda biochips based on droplet-aliquot operations and predictive analysis. In *Proceedings of of ICCD*, pages 615–622, 2017.

Index

Printed in the United States
by Baker & Taylor Publisher Services